POLLUTION SCIENCE, TECHNOLOGY AND ABATEMENT SERIES

ENFORCING FEDERAL POLLUTION CONTROL LAWS

POLLUTION SCIENCE, TECHNOLOGY AND ABATEMENT SERIES

Soil Remediation
Lukas Aachen and Paul Eichmann (Editors)
2009. ISBN 978-1-60741-074-4

Soil Remediation
Lukas Aachen and Paul Eichmann (Editors)
2009. ISBN: 978-1-60876-651-2 (Online Book)

Prediction of Performance and Pollutant Emission from Pulverized Coal Utility Boilers
N. Spitz, R. Saveliev, E. Korytni, M. Perelman, E. Bar-Ziv and B. Chudnovsky
2009. ISBN 978-1-60741-184-0

Fly Ash: Reuse, Environmental Problems and Related Issues
Peter H. Telone (Editor)
2009. ISBN: 978-1-60741-632-6

Enforcing Federal Pollution Control Laws
Norbert Forgács (Editor)
2010. ISBN: 978-1-60876-082-4

Biofouling: Types, Impact and Anti-Fouling
Jun Chan and Shing Wong (Editors)
2010. ISBN: 978-1-60876-501-0

Impact, Monitoring and Management of Environmental Pollution
Ahmed El Nemr (Editor)
2010. ISBN: 978-1-60876-487-7

POLLUTION SCIENCE, TECHNOLOGY AND ABATEMENT SERIES

ENFORCING FEDERAL POLLUTION CONTROL LAWS

NORBERT FORGÁCS
EDITOR

Nova Science Publishers, Inc.

New York

Copyright © 2010 by Nova Science Publishers, Inc.

All rights reserved. No part of this book may be reproduced, stored in a retrieval system or transmitted in any form or by any means: electronic, electrostatic, magnetic, tape, mechanical photocopying, recording or otherwise without the written permission of the Publisher.

For permission to use material from this book please contact us:
Telephone 631-231-7269; Fax 631-231-8175
Web Site: http://www.novapublishers.com

NOTICE TO THE READER

The Publisher has taken reasonable care in the preparation of this book, but makes no expressed or implied warranty of any kind and assumes no responsibility for any errors or omissions. No liability is assumed for incidental or consequential damages in connection with or arising out of information contained in this book. The Publisher shall not be liable for any special, consequential, or exemplary damages resulting, in whole or in part, from the readers' use of, or reliance upon, this material. Any parts of this book based on government reports are so indicated and copyright is claimed for those parts to the extent applicable to compilations of such works.

Independent verification should be sought for any data, advice or recommendations contained in this book. In addition, no responsibility is assumed by the publisher for any injury and/or damage to persons or property arising from any methods, products, instructions, ideas or otherwise contained in this publication.

This publication is designed to provide accurate and authoritative information with regard to the subject matter covered herein. It is sold with the clear understanding that the Publisher is not engaged in rendering legal or any other professional services. If legal or any other expert assistance is required, the services of a competent person should be sought. FROM A DECLARATION OF PARTICIPANTS JOINTLY ADOPTED BY A COMMITTEE OF THE AMERICAN BAR ASSOCIATION AND A COMMITTEE OF PUBLISHERS.

LIBRARY OF CONGRESS CATALOGING-IN-PUBLICATION DATA

Enforcing federal pollution control laws / editor, Norbert Forgács.
   p. cm.
 Includes index.
 ISBN 978-1-60876-082-4 (hardcover)
 1. Environmental law--United States. 2. Offenses against the environment--United States. 3. Law enforcement--United States. I. Forgács, Norbert.
 KF3775.E464 2010
 344.7304'632--dc22
         2009036566

Published by Nova Science Publishers, Inc. † New York

CONTENTS

Preface		**vii**
Chapter 1	DOJ Environment Litigation Accomplishments 2008 *United States Department of Justice*	**1**
Chapter 2	Environmental Enforcement: EPA Needs to Improve the Accuracy and Transparency of Measures Used to Report on Program Effectiveness *U. S. Government Accountability Office*	**33**
Chapter 3	Environmental Protection: EPA-State Enforcement Partnership Has Improved, but EPA's Oversight Needs Further Enhancement *U. S. Government Accountability Office*	**51**
Chapter 4	Environmental Compliance and Enforcement: EPA's Effort to Improve and Make More Consistent Its Compliance and Enforcement Activities *John B. Stephenson*	**85**
Chapter 5	Federal Pollution Control Laws: How are They Enforced? *Robert Esworthy*	**97**
Chapter 6	OECA FY 2008 Accomplishments Report: Protecting Public Health and Environment *United States Environmental Protection Agency*	**137**
Index		**167**

PREFACE

As part of its mission to protect human health and the environment, the Environmental Protection Agency's (EPA) enforcement office maintains civil and criminal enforcement programs to help enforce the requirements of major federal environmental laws such as the Clean Air Act and Clean Water Act. EPA's civil and criminal enforcement programs work with the Department of Justice (DOJ), and in some cases States, to take legal actions to bring polluters into compliance with federal laws. While civil enforcement actions require polluters to pay penalties and take other corrective actions, criminal enforcement actions may also include imprisonment. EPA estimates that these efforts achieved commitments to reduce 3.9 billion pounds of pollutants in the environment, primarily from air and water. EPA also assessed more than $195 million in civil and criminal fines and restitution during FY2008. Nevertheless, noncompliance with federal pollution control laws remains a continuing concern. This book focuses on the EPA's vow to pursue civil and criminal violations of environmental laws. This book consists of public documents which have been located, gathered, combined, reformatted, and enhanced with a subject index, selectively edited and bound to provide easy access.

Chapter 1 - I am proud to present this summary of the Environment and Natural Resources Division's litigation accomplishments for Fiscal Year 2008. This has been an outstanding year, in keeping with the Division's exemplary record of safeguarding the country's environment and natural resources. The Division handles cases involving more than 150 different statutes. We have a docket of more than 7,500 active cases and matters at every level of the federal court system as well as state courts. The Division both brings affirmative civil and criminal enforcement actions and defends federal agencies when their actions or decisions are challenged on the basis of our environmental or public lands and resources laws. As in past years, the Division achieved significant victories for the American people in each of the many areas for which it has responsibility. These responsibilities include protecting the nation's air, water, land, wildlife and natural resources, upholding our trust responsibilities to American Indians, acquiring needed lands for federal agencies, and otherwise defending important federal programs.

Chapter 2 - As part of its mission to protect human health and the environment, the Environmental Protection Agency's (EPA) enforcement office maintains civil and criminal enforcement programs to help enforce the requirements of major federal environmental laws such as the Clean Air Act and the Clean Water Act. EPA's civil and criminal enforcement programs work with the Department of Justice (DOJ), and in some cases states, to take legal actions to bring polluters into compliance with federal laws. While civil enforcement actions require polluters to pay penalties and take other corrective actions, criminal enforcement actions also may include imprisonment. EPA's enforcement office sets national priorities to focus resources on significant environmental risks and non-compliance patterns; prepares nationally significant civil and criminal cases for legal action by DOJ; uses 10 regional offices to implement civil enforcement actions on a day-to-day basis; and pursues criminal violations of environmental laws through its criminal enforcement office. The agency exercises its authority to independently pursue some violators through administrative proceedings—civil administrative actions—and to refer significant matters to DOJ when it believes cases need to be filed in federal court as civil judicial actions.[1] DOJ is responsible for prosecuting and settling civil judicial and criminal enforcement cases.

Chapter 3 - The Environmental Protection Agency (EPA) enforces the nation's environmental laws through its Office of Enforcement and Compliance Assurance (OECA). OECA sets overall enforcement policies and through its 10 regions oversees state agencies authorized to implement environmental programs consistent with federal requirements. GAO was asked to (1) identify trends in federal resources to regions and states for enforcement between 1997 and 2006, and determine regions' and states' views on the adequacy of these resources; (2) determine EPA's progress in improving priority setting and enforcement planning with states; and (3) examine EPA's efforts to improve oversight of states' enforcement programs and identify additional actions EPA could take to ensure more consistent state performance and oversight. GAO examined information from all 10 regions and 10 authorized states, among other things.

Chapter 4 - The Environmental Protection Agency (EPA) enforces the nation's environmental laws and regulations through its Office of Enforcement and Compliance Assurance (OECA). While OECA provides overall direction on enforcement policies and occasionally takes direct enforcement action, many enforcement responsibilities are carried out by EPA's 10 regional offices. In addition, these offices oversee the enforcement programs of state agencies that have been delegated the authority to enforce federal environmental protection regulations.

This testimony is based on GAO's reports on EPA's enforcement activities issued over the past several years and on observations from ongoing work that is being performed at the request of this Committee and the Subcommittee on Interior, Environment and Related Agencies, House Committee on Appropriations. GAO's previous reports examined the (1) consistency among EPA regions in carrying out enforcement activities, (2) factors that contribute to any inconsistency, and (3) EPA's actions to address these factors. Our current work examines how EPA, in consultation with regions and states, sets priorities for compliance and enforcement and how the agency and states determine respective compliance and enforcement roles and responsibilities and allocate resources for these purposes.

Chapter 5 - As a result of enforcement actions and settlements for noncompliance with federal pollution control requirements, the U.S. Environmental Protection Agency (EPA) reported that, for FY2008, regulated entities committed to invest an estimated $11.8 billion for judicially mandated controls and cleanup, and for implementing mutually agreed upon (supplemental) environmentally beneficial projects. EPA estimates that these efforts achieved commitments to reduce 3.9 billion pounds of pollutants in the environment, primarily from air and water. EPA also assessed more than $195 million in civil and criminal fines and restitution during FY2008. Nevertheless, noncompliance with federal pollution control laws remains a continuing concern. The overall effectiveness of the current enforcement organizational framework, the balance between state autonomy and federal oversight, and the adequacy of funding are long-standing congressional concerns.

This chapter provides an overview of the statutory framework, key players, infrastructure, resources, tools, and operations associated with enforcement and compliance of the major pollution control laws and regulations administered by EPA. It also outlines the roles of federal (including regional offices) and state regulators, as well as the regulated community. Understanding the many facets of how all federal pollution control laws are enforced, and the responsible parties involved, can be challenging. Enforcement of the considerable body of these laws involves a complex framework and organizational setting.

Chapter 6 - The strength of EPA's enforcement program is illustrated by an unprecedented run of record results. EPA holds polluters accountable. In FY 2008, EPA concluded civil and criminal enforcement actions requiring polluters to spend an estimated $11.8 billion, an agency record, on pollution controls, cleanup and environmental projects. This exceeds the FY 2007 amount by approximately $800 million.* This means that each workday OECA was securing agreements from violators to invest an estimated $47 million to achieve compliance. The combined total for the last five years is an estimated $45 billion ($5.5, $11.3, $5.4, $11.0, and $11.8 billion, respectively)—exceeding EPA's total annual budget over the same period.

In: Enforcing Federal Pollution Control Laws
Editor: Norbert Forgács

ISBN: 978-1-60876-082-4
© 2010 Nova Science Publishers, Inc.

Chapter 1

DOJ ENVIRONMENT LITIGATION ACCOMPLISHMENTS 2008[*]

United States Department of Justice
Environmental and Natural Resources Division

FOREWORD

I am proud to present this summary of the Environment and Natural Resources Division's litigation accomplishments for Fiscal Year 2008. This has been an outstanding year, in keeping with the Division's exemplary record of safeguarding the country's environment and natural resources. The Division handles cases involving more than 150 different statutes. We have a docket of more than 7,500 active cases and matters at every level of the federal court system as well as state courts. The Division both brings affirmative civil and criminal enforcement actions and defends federal agencies when their actions or decisions are challenged on the basis of our environmental or public lands and resources laws. As in past years, the Division achieved significant victories for the American people in each of the many areas for which it has responsibility. These responsibilities include protecting the nation's air, water, land, wildlife and natural resources, upholding our trust responsibilities to American Indians, acquiring needed lands for federal agencies, and otherwise defending important federal programs.

Turning to overall enforcement results, I am pleased to report that in Fiscal Year 2008 the Division secured nearly $115 million in civil penalties and $9.3 billion in corrective measures through court orders and settlements. In addition, the Division concluded 58 criminal cases against 108 defendants, obtaining 21 years and 3 months of jail time and nearly $52 million in fines.

Fiscal Year 2008 also featured some "firsts" in the history of federal civil and criminal enforcement.

---

[*] This is an edited, reformatted and augmented version of a U. S. Department of Justice publication dated 2008.

- The Largest Ever Single Environmental Settlement. As I previewed in the foreword to last year's accomplishments report, the Division concluded the largest environmental settlement in history when the court this year entered the final consent decree in United States v. American Electric Power (AEP), resolving claims under the Clean Air Act's new source review/prevention of significant deterioration provisions. Under the decree's terms, AEP will install and operate $4.7 billion worth of air pollution controls on 16 coal-fired power plants. When the consent decree is fully implemented, these air pollution controls and other measures will reduce air pollution by 813,000 tons a year compared with pre-settlement emissions, making this the largest reduction in air pollution achieved by any single settlement. AEP also paid a $15 million civil penalty and will spend $60 million on projects to mitigate the adverse effects of its past excess emissions. An unprecedented coalition of 8 states and 13 citizen groups joined the United States in the settlement.

- The Highest Superfund Cost Recovery. In United States v. W.R. Grace & Co., the Division recovered $252.7 million, the highest sum in the history of the Superfund program, in reimbursement of the United States' costs in connection with the cleanup of asbestos contamination in Libby, Montana.

- The Largest Civil Penalty for Clean Water Act Permit Violations. In United States v. Massey Energy Co., the Division obtained a civil penalty of $20,100,500, the largest civil penalty ever levied against a company for wastewater discharge permit violations. Massey, the fourth largest coal company in the United States, also agreed to take additional measures at its facilities nationwide to prevent an estimated 380 million pounds of sediment and other pollutants from entering the nation's waters each year. These compliance measures are unprecedented in the coal mining industry.

The Division also pressed forward this year with key environmental litigation initiatives:

Power Plant Enforcement. The Division has successfully litigated a number of significant Clean Air Act claims against operators of coal-fired electric power generating plants. The violations at issue arose from companies engaging in major life extension projects on their aging facilities without installing required pollution controls. To date, 15 of these matters have settled on terms that will result in reductions of nearly 2 million tons of $SO2$ and NOx each year once the $11 billion in required pollution controls are fully functioning.

Vessel Pollution Prosecutions. The Division, in partnership with U.S. Attorney's Offices across the country, continued its great success in the Vessel Pollution Initiative, a concentrated effort to prosecute those who illegally discharge pollutants from ships into oceans, coastal waters, and inland waterways or those who falsely document their activities. Over the past 10 years, the criminal penalties imposed in vessel pollution cases have totaled more than $200 million and responsible shipboard officers and shore-side officials have been sentenced to more than 17 years of incarceration. The initiative has resulted in a number of important criminal prosecutions of key segments of the commercial maritime industry, including cruise ships, container ships, tank vessels, and bulk cargo vessels. This year, the Division's appeal of *United States v. Jho* obtained the first appellate ruling on the scope of federal jurisdiction to prosecute log book offenses such as these and the meaning of the duty to maintain log books.

Addressing Air Pollution from Oil Refineries. The Division has made significant progress in combating Clean Air Act violations within the petroleum refining industry. To date, the Division's petroleum refinery enforcement initiative has produced settlements or other court orders that have addressed more than 96 individual refineries and 87% of the nation's refining capacity, and will reduce air pollutants by more than 331,000 tons a year.

Ensuring the Integrity of Municipal Wastewater Treatment Systems. The Division continued its efforts to protect the nation's waterways, by using the Clean Water Act to ensure the proper operation of municipal sewer systems. Since January 2006, courts have entered more than a dozen settlements in these cases, requiring long-term control measures estimated to cost in excess of $5 billion. The settlements the Division reached in Fiscal Year 2008 will ultimately reduce the volume of untreated sewage discharged into our waterways by tens of billions of gallons. This year, for example, the Division concluded a final, comprehensive settlement with the City of San Diego, resolving our action against the city stemming from unlawful discharges of sewage from its sewer system. Two previous decrees required the city to undertake interim measures at an estimated cost of $274 million. The third and final consent decree will require the city to continue to undertake capital projects and perform operations and maintenance through 2013, at a cost of an additional $1 billion, to prevent future spills of sewage from its system.

Record-breaking enforcement cases such as these, however, are only a part of the Division's work. A significant aspect of our case docket involves defending vital federal programs, including military and national security programs or taking affirmative action to support such programs. This year, working with several U.S. Attorney's Offices, the Division initiated almost 400 eminent domain cases referred by the Department of Homeland Security to acquire permanent interests in privately-owned lands along the United States/Mexico border needed to meet a Congressional mandate for secure fence construction. We also represented the Navy in several cases that challenged the Navy's use of mid-frequency active sonar throughout the world and in specific training exercises off the coast of California and Hawaii, as well as its use of low-frequency sonar, a new technology for anti-submarine warfare that is still in the experimental phase. These cases are critically important to the nation's security and military readiness, and our efforts culminated in success before the United States Supreme Court. On November 12, 2008, the Supreme Court, in *Winter v. NRDC*, reversed the Ninth Circuit's affirmance of a district court's preliminary injunction that imposed conditions on the Navy's use of sonar in training exercises in the Southern California Operating Area of the Pacific Ocean. The Supreme Court held that the lower courts had not given adequate weight to the harm the Navy said its training would suffer from certain of the conditions and that the lower courts had improperly assessed the equities and the public interest.

The Division promotes responsible stewardship of our natural resources by defending federal agencies charged with such tasks as determining whether a species should be listed as endangered or threatened, managing fishery resources in a way that balances various interests,

United States Department of Justice

overseeing water conservation projects, managing activities on federal lands that range from grazing to oil and gas leasing, and protecting the nation's forests from the risks of wildfire. This year, the Division had important successes in facilitating the work agencies do in all these areas.

The Division's work also secured critical water rights for the United States. In Fiscal Year 2008, the Division reached important settlements and secured favorable judgments ensuring access to the water necessary to maintain the vitality of natural resources and the water needed to support varied uses of the public lands, national forests, national parks, wildlife refuges, wild and scenic rivers, military bases, and federal reclamation projects throughout the West. As but one example, I joined the Secretary of the Interior, members of Congress, and representatives of the States of California and Nevada and the Pyramid Lake Paiute Tribe, among others, to sign an historic Truckee River Operating Agreement ("TROA"), the culmination of 15 years of negotiations the Division led. In addition to enhancing drought protection for the Cities of Reno and Sparks and securing Congressional approval of the interstate allocation between Nevada and California of the waters of Lake Tahoe, and the Truckee and Carson Rivers, TROA provides significant environmental benefits through more flexible and coordinated reservoir operations. This flexibility allows water to be stored and released for the benefit of threatened and endangered fish species in Pyramid Lake, water quality in the lower Truckee River, and instream flows on the Truckee River and tributaries under the California Guidelines.

The Division also remains at the forefront in carrying out the United States' trust responsibility to Indian tribes and resolving issues pertaining to American Indians. We work to protect tribal fishing and water rights, most notably this year in the case of *United States v. Oregon*. Many years ago, the United States prevailed in establishing the treaty fishing rights of four Columbia River Basin tribes. The taking of those fish, however, impacts anadromous species that are listed pursuant to the Endangered Species Act. The parties, after a decade of negotiations, concluded the Management Agreement for Fish Harvests on the Columbia and Snake Rivers in Washington, Oregon and Idaho for 2008-2017, consistent with the Endangered Species Act. This plan will improve fish habitat and allow the tribes to increase their catch as the populations of threatened species increase.

I would like to close on a personal note. I've been with the Department of Justice, in various capacities, nearly 12 years and served in the position of Assistant Attorney General for more than 18 months. The Department is a place populated by dedicated public servants, committed to the high ideals of public service, and each has sworn an oath to see that the laws are well and faithfully carried out. Those in the Environment and Natural Resources Division do so, without exception, in a way that should make our citizens proud, and most certainly in a way that has made me proud to be their colleague. I extend to them my congratulations for this year's many accomplishments and offer them my best wishes for the future. It bears repeating that the Division's unparalleled service and commitment show what a powerful difference our national government makes in the lives of those we serve.

Ronald J. Tenpas
Assistant Attorney General
Environment and Natural Resources Division
January 2009

Protecting Our Nation's Air, Land, and Water

Reducing Air Pollution from Power Plants

During the past year, the Division continued to successfully litigate Clean Air Act (CAA) claims against operators of coal-fired electric power generating plants. The violations arose from companies engaging in major life extension projects on aging facilities without installing required state-of-the-art pollution controls, resulting in excess air pollution that has degraded forests, damaged waterways, contaminated reservoirs, and adversely affected the health of the elderly, the young, and asthma sufferers. To date, 15 of these matters have settled on terms that will result in reductions of nearly 2 million tons of $SO2$ and NOx each year once the $11 billion in required pollution controls are fully functioning.

In Fiscal Year 2008, the Division achieved the largest environmental settlement in history when the court entered the final consent decree in *United States v. American Electric Power* (AEP), resolving claims under the CAA's new source review/prevention of significant deterioration provisions. Under the decree's terms, AEP will install and operate $4.7 billion worth of air pollution controls on 16 coal-fired power plants. When the consent decree is fully implemented, these air pollution controls and other measures will reduce air pollution by 813,000 tons every year compared with pre-settlement emissions, making this the largest reduction in air pollution achieved by any single settlement. AEP also paid a $15 million civil penalty and will spend $60 million on projects to mitigate the adverse effects of its past excess emissions. An unprecedented coalition of 8 states and 13 citizen groups joined the United States in the settlement.

Addressing Air Pollution from Oil Refineries

The Division also made progress in its national initiative to combat CAA violations within the petroleum refining industry by obtaining a consent decree with Sinclair Oil. With this settlement, the Division's petroleum refinery enforcement initiative has produced settlements or other court orders that have addressed more than 96 individual refineries and 87% of the nation's refining capacity, and will reduce air pollutants by more than 331,000 tons a year.

Sinclair Oil agreed to spend more than $72 million for new and upgraded pollution controls to reduce air pollution from its 3 refineries. Under the terms of the consent decree, Sinclair will reduce annual NOx and $SO2$ emissions by 1,100 tons and 4,600 tons, respectively. Sinclair also agreed to pay a civil penalty of $2.45 million and spend $150,000 on supplemental environmental projects as part of the settlement.

Reducing Air Pollution from Mobile Sources

The Division obtained a consent decree from a Taiwanese manufacturer and three American corporations in *United States v. McCulloch*, resolving claims that the defendants failed to meet CAA standards. Under the consent decree, the companies agreed to pay a $2 million civil penalty for importing and distributing some 200,000 chainsaws that would emit a total of 268 tons of excess hydrocarbons into the environment over their lifetime.

Controlling Contaminated Storm Water Run-off

The Division also fought for cleaner water by enforcing Clean Water Act (CWA) provisions governing discharge of storm water, which contains pollutants such as suspended solids, lead, and copper.

The Division achieved settlements with four of the largest home builders in the country: Centex, KB Home, Pulte, and Richmond. Together, they agreed to pay $4.2 million in civil penalties and to implement compliance programs at construction sites in 34 states and the District of Columbia that will prevent 1.2 billion pounds of sediment from polluting our waterways each year. Pulte agreed to complete a supplemental environmental project at a minimum cost of $608,000. Home Depot settled its storm water violations, agreeing to pay a $1.3 million civil penalty for violations at more than 30 construction sites in 28 states where its stores were being built. Home Depot also agreed to implement a nationwide compliance program with several St. Louis-area developers.

Republic Services agreed to construct and operate a comprehensive remedy for the Sunrise Mountain Landfill and to pay a $1 million penalty to resolve violations of the CWA. The remedy will be designed to withstand a 200-year storm and is expected to cost $36 million. Upon completion, it will prevent the release of more than 14 million pounds of contaminants annually, including storm water pollutants.

Ensuring the Integrity of Municipal Wastewater Treatment Systems

Through its aggressive national enforcement program, the Division continued to protect the nation's waterways by ensuring the integrity of municipal wastewater treatment systems. Since January 2006, courts have entered more than a dozen settlements in these cases, requiring long-term control measures estimated to cost in excess of $5 billion. The settlements the Division reached this year will reduce the discharge of untreated sewage into our waterways by tens of billions of gallons.

The Division achieved a final, comprehensive settlement with the City of San Diego, resolving our CWA action against the city stemming from unlawful discharges of sewage from its sewer system. Two previous decrees required the city to undertake interim measures at an estimated cost of $274 million. The third and final consent decree will require the city to continue to undertake capital projects and perform operations and maintenance through 2013, at a cost of an additional $1 billion, to prevent future spills of sewage from its system.

The Division also achieved an interim settlement with the City and County of Honolulu (CCH) that will correct the most significant problems in Honolulu's wastewater collection system. Under the terms of the consent decree, CCH will implement $300 million in projects. The United States and the State of Hawaii are continuing to work with CCH to resolve its remaining wastewater collection and treatment problems.

Protecting the Nation's Waters and Wetlands

The Division obtained a number of favorable settlements in enforcement actions to protect the nation's waters and wetlands from illegal fill. In *United States v. Johnson*, the Division sued an Arizona land developer and a contractor for violations of the CWA in bulldozing, filling, and diverting approximately five miles of the lower Santa Cruz River and a major tributary, the Los Robles Wash, without a permit from the Army Corps of Engineers. We negotiated a consent decree which, when entered, will require the defendants to pay a combined $1.25 million civil penalty, one of the largest penalties in the Environmental Protection Agency's (EPA) history under Section 404 of the CWA, which protects against the unauthorized filling of federally protected waterways.

The Division also negotiated a favorable settlement of *United States v. Sea Bay Development Corp.*, resolving allegations that the defendants discharged dredged or fill material into wetlands at an approximately 1,560-acre property in Chesapeake, Virginia, without a permit. Under several consent decrees, the defendants will pay civil penalties totaling $100,000. The consent decree with the site owner requires comprehensive restoration and mitigation on approximately 873 acres of the wetlands, which will be preserved in perpetuity under a conservation easement or deed restriction.

United States v. Fabian is a CWA civil enforcement action for the unauthorized filling of wetlands located along the Little Calumet River in Indiana. In 2007, the court granted our motion for summary judgment on liability. We subsequently resolved the remedy through a consent decree, under which the defendant will pay a small civil penalty and convey approximately 93 small parcels of real estate, including the site where the filling of wetlands occurred, into a trust. The decree then obligates the trust to effectuate restoration and off-site mitigation to the maximum extent allowable by the trust assets.

Reducing Air and Water Pollution at Other Diverse Facilities

The Division improved the nation's air and water quality by concluding regulatory enforcement actions against a variety of other facilities in diverse industries. In total, the Division obtained recoveries valued at more than $8.7 billion in injunctive relief; more than $105 million in civil penalties; and $25.4 million in supplemental environmental projects. One significant case was *United States v. Massey Energy Co.* There, the Division obtained the largest civil penalty ever levied against a company for wastewater discharge permit violations when Massey agreed to pay a $20,100,500 civil penalty. Massey, the fourth largest coal company in the United States, also agreed to take additional measures at its facilities nationwide to prevent an estimated 380 million pounds of sediment and other pollutants from entering the nation's waters each year. These compliance measures are unprecedented in the coal mining industry.

Protecting the Public against Vinyl Chloride

The Division has begun taking enforcement actions against manufacturers of vinyl chloride, which EPA has classified as a Group A human carcinogen. Exposure to the chemical has been linked to adverse human health effects, including liver cancer, other liver ailments, and neurological disorders.

In *United States v Georgia Gulf* (GG), the court entered a consent decree resolving claims under five statutes, including the CWA, CAA, and the Resource Conservation and Recovery Act (RCRA), arising out of violations at GG's facility in Aberdeen, Mississippi, a plant that manufactures PVC from vinyl chloride monomer. GG will pay a civil penalty of $610,000 and perform injunctive relief valued at $2.9 million.

Enhancing Pipeline Safety

The Division obtained a judgment on the merits following a five-week bench trial in *United States v. Apex Oil Co*. In entering judgment for the United States, the court directed Apex to perform substantial injunctive relief valued at more than $150 million. Apex had owned and operated a refinery and associated pipelines and sewers in Hartford, Illinois, from which releases of gasoline and other petroleum-based substances had contributed to a substantial subsurface plume of petroleum-based substances.

The Division secured additional relief in *United States v. Magellan Pipeline Co.* when the court entered a consent decree addressing 11 oil spills from Magellan's pipelines and other facilities. The consent decree requires Magellan to perform comprehensive injunctive relief valued at approximately $6.5 million to prevent future spills and to pay a civil penalty in the amount of $5.3 million.

ENSURING CLEANUP OF OIL AND HAZARDOUS WASTE

Conserving the Superfund by Securing Cleanups and Recovering Superfund Monies

The Division secured the commitment of responsible parties to clean up additional hazardous waste sites, at costs estimated in excess of $541 million, and recovered approximately $420 million for the Superfund to help finance future cleanups. Examples of some of the major Superfund cases resolved by the Division this year include: *United States v. Atl. Richfield Company* (the company agreed to pay $187 million to finance major cleanup along 120 miles of the Clark Fork River and other areas in southwestern Montana, with $103.7 million being available for remedial actions, $7.6 million to reimburse federal government for past costs, and $3.35 million for the federal government's natural resource damages (NRD); Atlantic Richfield will also pay the State of Montana an additional $72.5 million, which the state will use to finance additional natural resource restoration activities as part of the settlement); *United States v. City of Jacksonville* (the city agreed to clean up two Superfund sites at an estimated cost of $94 million); and *United States v. Exxon Mobil Corp.*

(101 defendants will ensure a site-wide $48 million cleanup of the Beede Waste Oil site in Plaistow, New Hampshire, pay more than $9 million for future federal and state oversight costs, and pay $17 million in past federal and state response costs).

Enforcing Clean-up Obligations in Bankruptcy Cases

The Division's bankruptcy practice has continued to grow, and this year we achieved notable success in several proceedings.

In *United States v. W.R. Grace & Co.*, the Division recovered $252.7 million, the highest sum in the history of the Superfund program, in reimbursement of the United States' costs in connection with the cleanup of asbestos contamination in Libby, Montana. The action settled a bankruptcy claim brought by the federal government to recover money for past and future costs of cleanup of contaminated schools, homes, and businesses in Libby. In 2003, after a 3-day trial, the federal district court in Montana awarded the United States more than $54 million for costs incurred by EPA through Dec. 31, 2001. That award was not paid due to W.R. Grace's filing for bankruptcy protection. The bankruptcy settlement resolved the 2003 judgment as well as continuing clean-up costs EPA has incurred since Dec. 31, 2001 and will incur in the future. EPA will place the settlement proceeds into a special account within the Superfund that will be used to finance future clean-up work at the site.

In *United States v. Apache Energy & Minerals Co.*, the court entered two consent decrees resolving three defendants' Comprehensive Environmental Response, Compensation, and Liability Act (CERCLA) liability at the California Gulch Superfund Site in Leadville, Colorado. The first, with Asarco LLC, resolves the United States' allowed unsecured claim in the Asarco bankruptcy proceeding by requiring the payment of $9.3 million for response costs, and $10 million for NRD. The second, with Newmont USA Limited and Resurrection Mining Company, requires these defendants to pay $8.5 million in response costs and $10.5 million for NRD, and to pay future oversight costs. Newmont and Resurrection are additionally required to undertake work to address the discharge of acid mine drainage at the site, at an estimated cost of $93 million.

In one of the most challenging bankruptcy proceedings, *In re: Asarco LLC*, the United States has continued to litigate and reach settlements on our claims for clean-up work and NRD at more than 50 sites, and is engaged in mediation in an effort to reach agreement on a plan of reorganization for the company. Asarco LLC, and its predecessor companies, operated in the mining, milling, and smelting industries for over 100 years. This left a legacy of environmental contamination in over 16 states. The bankruptcy, which was filed in 2005, is the largest environmental bankruptcy in history both in terms of the number of sites where Asarco is liable (approximately 80) and the total amount of Asarco's liability at those sites. The environmental claims and liabilities asserted against Asarco in the bankruptcy by the United States and the states exceed $2 billion.

Defending the Constitutionality of the Superfund Law

In addition to its enforcement actions to secure the cleanup of hazardous waste sites, the Division has also successfully defended lawsuits aimed at interfering with clean-up actions by EPA and other federal agencies. For example, in *Goodrich Corp. v. EPA*, Goodrich brought a complaint alleging that EPA has engaged in an unconstitutional "pattern and practice" of issuing unilateral administrative orders under CERCLA. In 2007, the court held that the statutory regime on its face satisfies due process requirements; however, the court initially allowed Goodrich to file an amended complaint challenging EPA's "pattern and practice" of administering the statute. In December 2007, the court dismissed the due process claim against EPA with prejudice, concluding that Section 113(h) of CERCLA deprived the court of subject matter jurisdiction over Goodrich's "pattern and practice" due process claim.

Protecting the Government against Unwarranted Claims of Clean-Up Liability

The United States filed a counterclaim to recover response costs incurred by EPA in *Raytheon Aircraft Co. v. United States*, a cost recovery action under CERCLA involving the Tri-County Airport in Herington, Kansas, which was owned and operated by the Army Air Corps during World War II. Following a 10-day bench trial, the court found the United States not liable for any portion of the clean-up costs incurred by Raytheon (approximately $6 million) and found Raytheon liable for EPA's clean-up costs (approximately $3 million), with prejudgment interest accruing since May 2000.

Protecting the Public Fisc against Excessive Clean-Up Claims

In *Basic Management, Inc. v. United States*, a contribution action under CERCLA concerning a former magnesium plant in Henderson, Nevada, owned by the United States during World War II, the court ruled that the plaintiff cannot recover any response costs paid by its insurance company and also declined to extend to CERCLA cases the collateral source rule that precludes defendants in tort cases from offsetting their tort liability with insurance recoveries. The effect of this ruling is to reduce the total of potentially recoverable costs to be allocated at trial from a claimed $70,000,000 to less than $1 million.

In CERCLA contribution cases in which the United States is a liable party, we have obtained favorable settlements that ensure the United States will not pay more than its fair share of clean-up costs. These include such multi-million dollar settlements in Fiscal Year 2008 as *E.I. DuPont de Nemours & Co. v. United States*; *Southern California Gas Co. v. United States*; *United States v. Albemarle Electric Membership Corp.*; *CBS Corp. v. United States*; *BASF Catalysts LLC v. United States*; *Lewis Operating Corp. v. United States*; and *Hercules v. United States*.

In *State of Ohio v. DOE*, a case involving claims for NRD under CERCLA in connection with the release of hazardous and radioactive material at the Department of Energy's (DOE) Feed Material Production Center, a former uranium processing facility in Fernald, Ohio, the

Division completed negotiation of a consent decree requiring DOE to pay $13.75 million in NRD and assessment costs, record an environmental covenant restricting most types of future development at the site, and implement a natural resource restoration plan committing DOE to maintain a series of natural resource restoration projects at the site.

PROMOTING RESPONSIBLE STEWARDSHIP OF AMERICA'S NATURAL RESOURCES AND WILDLIFE

Defending Endangered Species Act Listings and the Critical Habitat Program

The Endangered Species Act (ESA) requires either the United States Fish and Wildlife Service (FWS) or the National Marine Fisheries Service (NMFS), depending on the species, to determine whether a species should be listed as endangered or threatened under a set of five criteria and to designate critical habitat for listed species. In Fiscal Year 2008, we had notable success defending such determinations.

In *Arizona Cattle Growers' Ass'n v. Kempthorne*, the court upheld the FWS designation of critical habitat for the Mexican spotted owl against a variety of challenges. In *Marincovich v. Lautenbacher*, the court agreed with the Division that the factual and scientific determinations supporting the NMFS listing of the Lower Columbia River coho were rational and entitled to deference. The Division prevailed in *Home Builders v. FWS*, where the court upheld the FWS listing of the central California population of tiger salamander, concluding, among other points, that there was a rational connection between the threats to the species and the determination that it should be listed as threatened and that the Service had properly considered historical habitat loss. In *Defenders of Wildlife v. Kempthorne*, the court determined that the FWS decision not to list the Florida black bear because existing regulatory mechanisms were adequate was reasonable and supported by the administrative record in the case. Finally, in *Sierra Forest Products v. Kempthorne*, the court upheld the FWS determination that listing the West Coast distinct population of the Pacific fisher was warranted, but precluded by higher priority listing actions.

Defending NMFS Ocean Harvest Management

The Service is charged, under the Magnuson-Stevens Fishery Conservation and Management Act, with the difficult task of managing ocean commercial fishing not only to provide for conservation and sustainable fishing, but also to optimize yield. In several cases, the Division successfully defended the Service's balancing of these objectives. In *S. Offshore Fishing Ass'n v. Gutierrez*, the court upheld the emergency closure of a shark fishery where NMFS determined that overharvesting had occurred. Similarly, in *Starbound LLC v. Gutierrez*, the court upheld the Service's emergency rule prohibiting new entry into the 2007 Pacific Whiting Fishery where NMFS forecast overfishing if new entries were allowed. The Division prevailed in *North Carolina Fisheries Ass'n v. Gutierrez*, which upheld Service establishment of measures to end overfishing of certain snapper species.

Restoring the Florida Everglades

The Division continued to contribute to the restoration and protection of the Everglades ecosystem – including the 1.3 million-acre Everglades National Park, the largest, most important subtropical wilderness in North America – by acquiring lands within Everglades National Park and the Big Cypress National Preserve, as well as lands critical to the Army Corps of Engineers' project to improve water deliveries in the area.

Maintaining an Appropriate Management Regime for the Missouri River System

The Army Corps of Engineers manages the Missouri River System (consisting of six dams and reservoirs) for a variety of overlapping purposes, such as navigation, flood control, irrigation, and hydropower. In order to ensure that water resource decisions best serve these varied needs, the Corps issues a Master Manual that describes its water control plan. In 2006, the Corps made changes to the Master Manual to protect the endangered pallid sturgeon. In *State of Missouri v. United States Army Corps of Eng'rs*, the State of Missouri sued the Corps, alleging that it had violated the National Environmental Policy Act (NEPA) by using the Master Manual to provide a spring pulse of water for aquatic habitat maintenance without fully evaluating the environmental impacts. The Division successfully explained the Corps' actions and secured sound precedent that permits the Corps to balance water, navigation, energy, and agricultural needs, as well as protect endangered species.

Facilitating Dredging Projects Critical for Commerce

The Port of New York and New Jersey must be dredged to maintain navigation and commerce estimated to generate about $200 billion annually in direct and indirect benefits. Due to past and present pollution, managing dredged material from the port has become increasingly difficult. *NRDC v. United States Army Corps of Eng'rs* challenged the environmental analysis of a Corps' project to deepen the navigational channels of the New York/New Jersey Harbor. While the court found the Corps' analysis to be partly inadequate, the court agreed that the dredging should not be enjoined. The Division ultimately achieved a favorable settlement that allowed essential dredging to continue while the Corps provided additional environmental analysis of the ongoing project. In *Jones v. Rose*, the Division also defended NEPA claims raised against maintenance dredging and channel deepening in Portland, Oregon. The court granted summary judgment for the Corps on all counts.

Litigating Federal Land Management Programs and Policies

The federal government manages vast swaths of land for a variety of purposes, some well-known (such as outdoor recreation) and others less well-known (such as pest control).

Land management programs and policies are sometimes unpopular with one or more user groups and become the subject of litigation.

The Division is successfully defending policies related to the Sierra Nevada Forest Plan Amendment (Framework) governing 11.5 million acres in 11 national forests in the Sierra Nevada region of California. These policies include desperately needed fuel treatments to reduce the threats of catastrophic wildfire as well as to meet the habitat needs of species dependent on old growth forests. The Framework is the subject of four related lawsuits, *Sierra Forest Legacy v. Rey*; *California v. USDA*; *Pacific Rivers Council v. United States Forest Service*; *and California Forestry Ass'n v. Bosworth*. Recently, the court issued a decision in favor of the Forest Service on virtually all claims in the four lawsuits.

Other equally wide-ranging forest management activities of the Forest Service and the Bureau of Land Management have benefitted from the Division's legal efforts. For example, in *Lands Council v. McNair*, the Division prevailed against a challenge to the Mission Brush Project, a forest restoration project, on the Idaho Panhandle National Forest, under the National Forest Management Act (NFMA), NEPA, and the Administrative Procedure Act (APA). The project is designed to restore old-growth forest structures and address excessive stand density that is facilitating forest destruction by fire, insect infestation, and disease. The Ninth Circuit, in an *en banc* decision, concluded that it had erred in an earlier case by creating a requirement not found in any relevant statute or regulation and defying established law concerning the deference owed to agencies. The court disagreed that the Forest Service had violated the NFMA, found that the record supported the Forest Service's conclusions with regard to the effect of the project on species diversity, and approved the Forest Service's use of the amount of suitable habitat for a particular species as a proxy for the viability of that species. The court also rejected Lands Council's claim that the Forest Service violated NEPA by failing to include a discussion of the scientific uncertainty surrounding its strategy for maintaining species viability. Finally, the court held that the district court properly denied a preliminary injunction, noting that the standard for determining the appropriateness of a preliminary injunction does not allow a court to abandon a balance of harms analysis simply because a potential environmental injury is at issue.

In some instances, plaintiffs also challenge federal forest programs by seeking to invalidate ESA coverage for forest plans or projects. In Fiscal Year 2008, the Division was successful in defending against several such claims. In *Alliance for Wild Rockies v. United States Forest Service*, the court upheld a FWS and Forest Service determination that the challenged timber project was consistent with road density standards for grizzly bears. In *Western Watershed Project v. BLM*, the court upheld a broad programmatic biological opinion for a BLM resource management plan.

Ensuring the Limitations of Federal Jurisdiction Are Enforced

The APA and other special review provisions circumscribe federal jurisdiction, as do the requirements of standing and other jurisdictional prerequisites. This year, the Division successfully raised these defenses in appropriate cases. In *John R Sand & Gravel Co. v. United States*, the Department, through the Solicitor General's Office with the support of the Division, prevailed before the Supreme Court in its view that the statute of limitations for bringing claims for compensation against the United States in the Court of Federal Claims was jurisdictional and could not be waived by failure to assert it at the trial court level.

In *Coos County Bd. of County Comm'rs v. Kempthorne*, the lower and appellate courts agreed that the plaintiffs could not compel FWS to take action based on its five-year species status review because that internal process did not result in a final agency action subject to judicial review. Similarly, in *Am. Forest Resource Council v. Hall*, the court dismissed a challenge to the Service's five-year status review for the marbled murrelet.

The Division invoked the requirement to issue a 60-day notice of intent to sue under the ESA citizen suit provision, obtaining dismissal of claims in *Defenders of Conewango Creek v. Echo Developers*; *Nez Perce Tribe v. NOAA Fisheries*; and *Man Against Extinction v. Hall*.

We raised the defense of standing, obtaining dismissal of claims in *Center for Biological Diversity v. Kempthorne*, in which the plaintiffs sought relief under ESA regarding 50 species of foreign butterflies; *Glasser v. NMFS*, in which the plaintiff challenged a conservation plan, but did not allege injury with regard to the species covered by the plan; and *Miccosukee Tribe v. United States*, in which the tribe failed to allege that any of its members actually live in, hunt in, or otherwise use the area of Everglades National Park to be used for the challenged project.

The Division continued to obtain favorable rulings on the defense of mootness. In *Wilderness Soc'y v. Kane County*, the court found the case moot because Kane County had rescinded the challenged ordinance. In *Aquifer Guardians in Urban Areas v. FWS*, the court dismissed the case because the challenged transmission line was completed and the Army Corps of Engineers had verified compliance with ESA restrictions. In *National Parks and Conservation Ass'n v. United States Army Corps of Eng'rs*, claims against the challenged discharge permit and its ESA compliance were determined to be moot where the permit had expired, there was no existing authorization to declare invalid or set aside, and there was no authorized filling to enjoin.

Determining the Impact of the Religious Freedom Restoration Act on Land Management Agencies

In *Navajo Nation v. United States Forest Service*, four Indian tribes challenged the Forest Service's approval of an expansion of a ski resort on the San Francisco Peaks near Flagstaff, Arizona, under the Religious Freedom Restoration Act (RFRA). The expansion included snowmaking using reclaimed water, which in the view of Indian religious practitioners, desecrates the peak. The Ninth Circuit, in an *en banc* decision, held that the Forest Service's approval did not violate the RFRA because the proposal does not place a substantial burden on the plaintiffs' exercise of religion by forcing the plaintiffs to act contrary to their religion under the threat of a legal penalty or choose between their religion and the receipt of a government benefit.

CRIMINALLY ENFORCING OUR NATION'S POLLUTION AND WILDLIFE LAWS

Reducing Pollution from Ocean-Going Vessels

The Vessel Pollution Initiative is an ongoing, concentrated effort to detect, deter, and prosecute those who illegally discharge pollutants from ships into the oceans, coastal waters, and inland waterways, and those who falsely document their activities. Enforcement is primarily pursuant to the MARPOL Treaty, and its United States implementing legislation, the Act to Prevent Pollution from Ships (APPS), which requires vessels to maintain log books recording all transfers and discharges of oily waste. When a vessel pulls into a U.S. port and presents a false log book, the consequences are severe. Over the past 10 years, the criminal penalties imposed in vessel pollution cases have totaled more than $200,000,000 and responsible shipboard officers and shore-side officials have been sentenced to more than 17 years of incarceration. The initiative has resulted in a number of important criminal prosecutions of key segments of the commercial maritime industry, including cruise ships, container ships, tank vessels, and bulk cargo vessels.

This year, the Division continued to have great success prosecuting deliberate violations as the representative cases below illustrate.

The Division's appeal of *United States v. Jho* obtained the first appellate ruling on the scope of federal jurisdiction to prosecute log book offenses and the meaning of the duty to maintain them under APPS. Here, the indictment charged the defendants with the failure to "maintain" oil record books for the M/T Pacific Ruby, a foreign-flagged oil tanker that delivered petroleum products to ports along the United States' gulf coast; and alleged that the defendants failed to record unlawful discharges of petroleum-contaminated wastewater that occurred at sea. The Fifth Circuit reversed the district court's dismissal of the indictment, holding that the regulatory duty to "maintain" the record books is not limited to the duty to make correct entries when discharges occur, but includes the obligation to "ensure that [the record book] is accurate . . . upon entering the ports of navigable waters of the United States." The court further determined that there are no principles of international law that prevent the United States from prosecuting entry of U.S. ports with inaccurate record books as violations of domestic law in port; and that various articles of the United Nations Convention on the Law of the Sea were inapplicable to violations of domestic law committed in port.

In *United States v. Nat'l Navigation Co.* (NNC), the defendant, an Egyptian shipping operator, pled guilty to 15 felonies involving conduct aboard 6 vessels in NNC's fleet, including APPS and making false statements to federal officials. NNC was sentenced to pay a total penalty of $7.25 million – the largest ever in the Pacific Northwest for a case involving the falsification of ship logs to conceal deliberate pollution from ships. Of this amount, $2,025,000 will go toward funding community service projects. The company was also required to implement a comprehensive environmental compliance plan (ECP).

In *United States v. Mark Humphries*, the defendant was convicted for violating APPS, obstruction of an agency proceeding, and two false statement violations. Humphries was sentenced to serve six months' incarceration followed by a two-year term of probation. Humphries was a former chief engineer for the M/V Tanabata, a vessel managed by Pacific Gulf Marine (PGM). In 2007, PGM was sentenced to pay a $1 million fine, make a community service payment of $500,000, complete a 3-year term of probation, and implement an ECP.

Guilty pleas or convictions were also reached in four additional cases involving vessel operators and crew members in *United States v. B. Navi Ship Mgmt. Servs.; United States v. Reederei Karl Schlueter; United States v. Pacific Operators Offshore;* and *United States v. Ionia Mgmt. S.A.* These defendants were sentenced to pay a total of $7.55 million in fines and $700,000 in community service, with each individual serving a term of probation for crimes including APPS violations, false statements, obstruction of justice, and violations of the Outer Continental Shelf Lands Act.

Protecting the Country's Wetlands

In *United States v. Lucas*, the defendants sold house lots and installed septic systems in wetlands, while representing that the properties were dry. The septic systems failed, rendering the properties uninhabitable and causing sewage discharges into the wetlands and adjacent waters. A jury convicted Lucas and the other defendants for conspiracy, mail fraud, and violations of Sections 402 and 404 of the CWA. The defendants asserted that under the Supreme Court's 2006 decision in *Rapanos v. United States*, which was decided after the convictions, EPA and the Army Corps of Engineers lacked regulatory authority over their actions. The Fifth Circuit affirmed the convictions, finding that the evidence was sufficient to satisfy the CWA jurisdictional standards set forth in the plurality, concurring, or dissenting opinions in *Rapanos* and rejecting an unconstitutional vagueness challenge due to abundant evidence that defendants should have known the wetlands might be regulated.

Safeguarding Our Nation's Groundwater from Hazardous Waste Pollution

In *United States v. Dennis Pridemore*, the defendant, the former president and manager of Hydromex Inc., was charged with having illegally stored and disposed of hazardous waste including heavy metals cadmium, chromium, and lead that he had been paid to recycle into marketable products. Pridemore admitted that instead of doing so, he buried the wastes in trenches and produced faulty products that leached heavy metals into the surrounding soil and groundwater. He created false documents making it appear to regulators that he had customers for the products he claimed to be making and selling. Pridemore pled guilty to committing four RCRA violations, and to making two false statements. He was sentenced to serve 41 months' incarceration followed by a 3-year term of probation.

Securing and Protecting our Nation's Highways from the Illegal Transportation of Hazardous Waste

In *United States v. Krister Evertson*, the defendant arranged for the transportation of sodium metal and several above-ground storage tanks which contained sludge without proper shipping documents. Sodium metal and the materials in the tanks are highly explosive when mixed with water, and Everston failed to take protective measures to reduce the risk that the transported material would react and damage persons or property. Evertson was convicted of two RCRA storage and disposal violations and with violating the Hazardous Materials Transportation Safety Act. He was sentenced to serve 21 months' incarceration followed by a 3-year term of probation and to pay $421,049 in restitution for illegally transporting hazardous materials and illegally storing hazardous waste.

Protecting the Environment, Public Health, and Worker Safety

In *United States v. Spencer Environmental Inc.* (SEI), the corporate defendant pled guilty to a RCRA violation for accepting corrosive and ignitable hazardous wastes without a permit and to mishandling waste oil in violation of RCRA. SEI was sentenced to pay $150,000, half of which will go toward a community service payment. SEI's president, Donald Spencer, pled guilty to mishandling waste oil and was sentenced to serve six months' incarceration followed by a one-year term of probation.

In the *United States v. W.R. Grace & Co*, the Division obtained critical victories in several interlocutory appeals to the Ninth Circuit that will allow this important CAA case to go trial in early 2009. The company and several of its officers stand charged with conspiracy and substantive violations of the CAA for knowingly endangering the lives of workers at its vermiculite mine and of residents in the nearby town of Libby, Montana. The district court had entered a series of pre-trial rulings, in particular on the substantive elements of the CAA violations, that would have gutted the government's case.

Cracking Down on Illegal Asbestos Removal

In *United States v. Cleve-Allan George*, the defendant, who had been convicted in 2005 on 16 counts, including CAA and false statement violations, was sentenced to serve 33 months' incarceration, followed by a 3-year term of probation. George and his co-defendant did not follow asbestos work practice regulations, and filed false air monitoring reports related to a remediation project in a HUD-funded housing project.

In *United States v. Branko Lazic*, the defendant pled guilty to 1 CAA violation for the improper removal of asbestos at an elementary school, and was sentenced to serve 6 months' home confinement, followed by a 3-year term of probation, and to complete 50 hours of community service.

In *United States v. Robert Langill*, the defendant pled guilty to violating the CAA in connection with an illegal asbestos abatement at the U.S. Naval Air Station, Patuxent River, and was sentenced to serve 60 days' incarceration, followed by 10 months' home detention, and to serve a 2-year term of probation.

Safeguarding Our Fragile Ecosystem on the North Slope of Alaska

In *United States v. British Petroleum Exploration (Alaska), Inc. (BPXA)*, the corporate defendant failed to heed the many warning signs of imminent internal corrosion of oil transit lines that a reasonable operator should have recognized. This failure resulted in more than 200,000 gallons of crude oil on the North Slope spreading over 2 acres of tundra. BPXA's failure to allocate sufficient resources, due to cost-cutting measures, led to the failure of a section of the oil transit line which had not been inspected for eight years. BPXA pled guilty to a CWA violation for the largest-ever spill of crude oil on the north slope of Alaska. BPXA was sentenced to pay a $20 million fine, followed by a 3-year term of probation. Four million dollars will be used for research in support of the arctic environment in the State of Alaska on the North Slope, and $4 million in restitution will be paid to the State of Alaska. A second spill involving approximately 1,000 gallons of oil, that led to the shut down of Prudhoe Bay oil production on the eastern side of the field, was also covered by the plea agreement.

Safeguarding Aquatic Life and Water Quality in and around the Gulf of Mexico

In *United States v. Citgo Petroleum Group*, the defendant, who operated a Louisiana refinery, failed to maintain 2 storm water tanks and to build a planned third tank, which led to the discharge of 53,000 barrels of oil to the Calcasieu Estuary. The illegal discharge overran the company's storm water system resulting in limited commercial transportation on the waterways for approximately 10 days. Citgo pled guilty to a negligent violation of the CWA and was sentenced to pay a $13 million fine, the largest fine for a misdemeanor CWA violation. Additionally, the company implemented an ECP to ensure the estuary is protected from this kind of spill in the future.

In *United States v. Rowan Cos.*, the corporate defendant operated and cleaned offshore drilling rigs, creating substantial amounts of waste from routine maintance and sandblasting operations, including hydraulic oil, chemicals, paint, and other materials that were dumped directly into the Gulf of Mexico. Rowan pled guilty to three felonies and was sentenced to pay a $7 million dollar criminal fine along with $2 million in community service payments. In addition, the company will add an environmental division and implement an ECP to contain debris from future sandblasting operations. Nine supervisory employees of Rowan pled guilty and were fined and sentenced to terms of probation for their roles related to Rowan's violations.

Keeping Nuclear Power Safe

In *United States v. David Geisen*, the defendant was convicted of concealment and false writing violations for his role in a scheme to keep information from the Nuclear Regulatory Commission, and with making false statements to the Commission. Geisen was sentenced to serve four months' home detention. His co-defendant, Andrew Siemaszko, was convicted in a separate trial of three false statement violations and is awaiting sentencing.

Protecting Endangered Sea Turtles from International Smugglers

In *United States v. Esteban Lopez Estrada*, the Division achieved notable success in the investigation and prosecution of four wildlife smuggling rings – two based in Mexico and two in China – engaged in illegal trafficking in endangered or otherwise protected sea turtles and other protected species, and products made from their parts. The defendants bought and sold exotic leathers, including sea turtle, caiman, ostrich and lizard skins, and manufactured boots and belts from the skins to sell to customers in the United States. Other sea turtle parts were used to manufacture and sell guitar picks and violin bows. The investigation, known as "Operation Central," was a long-term undercover investigation run out of a store front in Denver, Colorado. Thus far, 7 defendants have pled guilty to charges including conspiracy, smuggling, and money laundering and have been sentenced to a total of 107 months of incarceration. Four indicted defendants remain at large.

Safeguarding U.S. Consumers from the False Labeling of Fish

In *United States v. True World Foods Chicago LLC*, True World imported fish commonly known as basa or Vietnamese catfish, but falsely labeled as sole, thereby reducing the amount of duty legally due. True World pled guilty to a Lacey Act violation for its role in purchasing and re-selling the falsely labeled frozen fish fillets and was sentenced to pay a $60,000 fine. The company further forfeited $197,930, which represented the purchase value of the fish. One company employee, David Wong, pled guilty to violating the Lacey Act for his role, while co-defendant, Henry Yip, pled guilty to a misbranding violation for his involvement in this scheme.

Prosecuting Illegal Hunting and Fishing

In *United States v. Eric Leon Butt, Jr., d/b/a Outdoor Adventures*, Butt, a Colorado big-game outfitter, pled guilty to conspiracy to violate the Lacey Act stemming from the interstate sale and transport of deer, elk, and black bear. Sentencing is pending. His co-defendant, Scott LeBlanc, pled guilty to a misdemeanor violation of the Lacey Act and was sentenced to pay a $3,000 fine, pay $5,000 in restitution, and complete a 2-year term of probation, and is banned from hunting in Colorado for 5 years. A third co-defendant, Paul Ray Weyand, pled guilty to three misdemeanor Lacey Act violations.

In *United States v. Zane Fennelly*, the defendant, the captain of a commercial fishing vessel, pled guilty to disposing of and attempting to destroy three bags containing spiny lobsters that were illegally caught within the exclusive economic zone of the United States. Fennelly dumped the illegally caught lobsters as the United States Coast Guard was approaching his vessel. Fennelly was sentenced to serve four months' incarceration followed by a one-year term of probation.

Defending Vital Federal Programs and Interests

Defending EPA's Air Pollution Standards for the Chemical Manufacturing Industry

NRDC v. EPA upheld EPA's final CAA rule setting National Emission Standards for Organic Hazardous Air Pollutants from the Synthetic Organic Chemical Manufacturing Industry. The court held that EPA could reasonably determine that a standard that reduced cancer risk to 1-in-10,000 was sufficient to satisfy the statute. Further, the agency could consider costs in assessing the "ample margin of safety" it was required to provide in the standard, given that Congress referenced in the statute a benzene standard in which EPA had also considered costs. Finally, the court upheld the record basis for EPA's decision, finding that EPA reasonably relied on industry-supplied data in reaching its decision and had reasonably responded to comments regarding the agency's alleged failure to regulate all sources of emissions of hazardous air pollutants in the category.

Upholding EPA's CAA Permitting Decisions and Enforcement Discretion

In *Citizens Against Ruining the Environment v. EPA*, the Seventh Circuit upheld EPA's decision not to object to Illinois' issuance of CAA Title V operating permits to six Midwest Generation power plants. The court first found that the Illinois Attorney General lacked standing, having failed to explain why the court should entertain a challenge by one state agency to a decision of a second agency of the same state. The court also ruled that EPA reasonably interpreted the CAA in deciding that petitioners had not "demonstrated" the permittee's noncompliance with the Act, where EPA had commenced judicial enforcement proceedings against the permittee, the permittee was contesting those allegations, and they had not been judicially resolved.

Similarly, in *Sierra Club v. EPA*, a petition for review of EPA's denial of a request to object to a CAA Title V permit for certain Georgia Power facilities, the Eleventh Circuit held that EPA acted reasonably and within its discretion in determining that a citizen petitioner had not made an adequate "demonstration" of a violation where the petition rested entirely on EPA's own allegations in an enforcement complaint that had not yet been adjudicated.

Protecting against Premature Challenges to Agency Action

In *State of California v. EPA*, the petitioners sought review of the EPA Administrator's letter to California Governor Schwarzenegger informing him of EPA's intent to deny California's request for a waiver of CAA preemption for California's proposed regulation of motor vehicle greenhouse gas emissions. EPA subsequently published a formal denial decision in the *Federal Register*. The court granted our motion to dismiss the challenge to EPA's letter, finding that the letter did not constitute reviewable final agency action.

Upholding EPA's Safe Drinking Water Regulations

In *City of Portland, Oregon v. EPA*, the court denied the petitioners' request to overturn EPA's National Primary Drinking Water Regulations setting requirements to reduce levels of cryptosporidium and other microbial pathogens in drinking water. The court found that EPA had adequately considered and addressed all issues during the rulemaking.

In another case involving regulation of drinking water, *Miami-Dade County v. EPA*, the petitioners sought review of a final rule entitled "Underground Injection Control Program -- Revision to the Federal Underground Injection Control Requirements for Class I Municipal Disposal Wells in Florida." The court found that EPA had reasonably addressed in the rule the risks posed by non-biological contaminants, pathogens, nutrients, and other contaminants, and that EPA permissibly compared the utility of underground injection with other forms of waste disposal. The court also held that EPA reasonably considered geologic variation in determining where the final rule would apply. Finally, the court upheld as reasonable EPA's resolution of a number of record-based issues, including findings EPA made in a risk assessment and its decision to require high-level disinfection before injection of wastes.

Defending against Liability for Damage Caused by Hurricane Katrina

In *Louisiana Envtl. Action Network v. United States Army Corps of Eng'rs*, the plaintiff sued the Corps of Engineers under RCRA, alleging the existence of an imminent and substantial endangerment and seeking to force the testing and any necessary cleanup of sediments deposited in the vicinity of the Pratt Drive breach of the London Avenue Canal following Hurricane Katrina. The court dismissed two complaints because RCRA does not abrogate the immunity that Congress conferred on the Corps in the Flood Control Act with regard to its flood control activities.

Defending Agency Post-September 11 Conduct

In *Benzman v. Whitman*, a case jointly handled by the Environment and Natural Resources and Civil Divisions, the plaintiffs who lived, worked, or went to school in southern Manhattan, alleged that they were injured by EPA official statements (including by then-Administrator Whitman) claiming air quality after the collapse of the World Trade Center was better than it actually was. As we argued, the Second Circuit held that the alleged actions by Whitman and EPA did not rise to a constitutional violation. The court noted the competing considerations to which both Whitman and EPA were subject, and held that imposing liability for actions taken under these circumstances could seriously hamper government decision-making. The court affirmed the dismissal of the claims, holding that the plaintiffs had failed to identify duties that EPA was required to carry out.

Defending Untimely Challenges to Agency Action

West Virginia Highlands Conservancy v. Johnson involved a citizen suit under RCRA, alleging that EPA had failed to perform a nondiscretionary duty to conduct a study on coal mining wastes, prepare a report to Congress, and make a determination whether the regulation of coal mining wastes as hazardous is warranted. The court granted our motion to dismiss the complaint as untimely on alternative grounds.

Enhancing the Nation's Energy Infrastructure

With the nation's growing imperative to efficiently and economically produce energy for its citizens and businesses, the Division's work in securing the rights-of-way and other requisites attendant to energy production and delivery has become increasingly important. For instance, the Division has been called upon to litigate disputes related to the designation of electricity transmission corridors in the Mid-Atlantic and the Southwest. In related cases, *National Wildlife Fed'n v. DOE*, and *Piedmont Envtl. Ctr. v. DOE*, we succeeded in having NEPA, National Historic Preservation Act, and Energy Policy Act claims challenging designation of the Mid-Atlantic Corridor dismissed.

Naturally, in order to provide oil and gas for energy production, it has to be located. The Division has defended suits over permits issued under the Outer Continental Shelf Lands Act to conduct seismic explorations in the Chukchi and Beaufort Seas off Alaska. After hearing the Division's arguments, the court found in *Native Village of Point Hope v. Minerals Management Service* that the agencies appropriately issued their authorizations on the basis of an environmental assessment and finding of no significant impact, even though they were still studying longer-term potential impacts. In *Center for Biological Diversity v. Kempthorne*, the Division successfully defended the FWS's issuance of a regulation and accompanying environmental assessment authorizing the incidental take of polar bears for five years during oil and gas operations in the Beaufort Sea. Of significance, the court concluded that the assessment adequately accounted for the combined effect of incidental take and climate change on polar bears. In yet another case, *North Slope Borough v. Minerals Management Service*, the Division presented persuasive evidence in support of an oil and gas lease sale in the Beaufort Sea challenged for its alleged impacts on polar bears, bowhead whales, and other wildlife.

Defending the Federal Highway Administration's Traffic Projects

As the nation grows more and more mobile and increases in population, it is increasingly important that our roads and bridges be safe, efficient, and minimize environmental impacts. The Division continues to play a significant role in defending the Federal Highway Administration's (FHWA) projects designed to address both traffic control and safety. This year saw some significant victories for the Division. In one example, the plaintiffs in the consolidated cases of *Audubon Naturalist Soc'y v. DOT* and *Environmental Defense v. DOT* challenged the Intercounty Connector highway project linking I-95/US 1 with I-270 across

Prince George's and Montgomery Counties in the Maryland suburbs outside Washington, D.C. This high profile project has been in the works for years to alleviate congestion outside of the I-495 beltway, which is among the worst in the nation. Along with the State of Maryland, we successfully defended against the myriad claims raised under NEPA, the Department of Transportation Act Section 4(f), the CWA, the CAA, and the Federal Aid to Highways Act.

The State of Indiana has been searching for over half of a century for a major highway route across the southwestern quadrant of the state. The FHWA, along with the Indiana Department of Transportation, took a hard look at the benefits and detriments of a particular route to complete that search. While acknowledging differences of opinion as to the project, the court in *Hoosier Envtl. Council v. DOT* agreed that the FHWA and the Indiana Department of Transportation had properly analyzed the effects of the proposed new I-69 segment from Indianapolis to Evansville designed to facilitate the international trade route from the Canadian border at Huron, Michigan, to the Mexican border at Laredo, Texas. The court also found that there were no violations of the CWA, ESA, or the Department of Transportation Act.

Securing Needed Water Rights for the United States

The Division is involved in litigating federal water rights claims for a variety of purposes. For example, the Great Sand Dunes National Park and Preserve Act of 2000 directs the Department of the Interior to obtain water rights under Colorado law to maintain groundwater levels under the Great Sand Dunes National Park and to protect the wetlands, streamflows, and other hydrology-dependent surface resources of the park for future generations. As a result of the Division's efforts, the Colorado water court granted the United States' claim and decreed to the National Park Service the rights to the groundwater underlying the park, which is critical to the look, the feel, and the ecology of the park.

In another water rights development this year, the Assistant Attorney General joined the Secretary of the Interior, members of Congress, and representatives of the States of California and Nevada and the Pyramid Lake Paiute Tribe, among others, to sign the historic Truckee River Operating Agreement ("TROA"), the culmination of 15 years of negotiations the Division led. Under the Truckee-Carson-Pyramid Lake Water Rights Settlement Act of 1990, the TROA is a prerequisite for Congressional approval of an allocation of the waters of Lake Tahoe, and the Truckee and Carson Rivers among the states, tribe, and stakeholders. In addition to enhancing drought protection for the Cities of Reno and Sparks, TROA provides significant environmental benefits through more flexible and coordinated reservoir operations. This flexibility allows water to be stored and released for the benefit of threatened and endangered fish species in Pyramid Lake, water quality in the lower Truckee River, and instream flows on the Truckee River and tributaries under the California Guidelines.

Defending Federal Agency Program Assessments of Impacts on Species

In a variety of cases, plaintiffs assert that federal agencies have not adequately considered their obligations under species protection laws when carrying out actions. In Fiscal Year 2008, the Division had favorable results in many such challenges. In *Nat'l Wildlife Fed'n v. Harvey*, the court held that the FWS reasonably concluded that an Army Corps of Engineers project would not adversely affect the newly rediscovered ivory-billed woodpecker. Similarly, in *Salmon Spawning and Recovery Alliance v. Lohn*, the court found that the NMFS biological opinion on the fishery management plans of Washington State and 17 Tribes for Puget Sound fisheries was reasonable. The Division successfully argued in *Center for Biological Diversity v. HUD* that HUD loan guarantees did not cause residential development requiring ESA consultation.

Acquiring Property for Public Purposes

The Division exercises the federal government's power of eminent domain to enable agencies to acquire land for various purposes. In this work, the Division is mindful of its goal to achieve results equitable to individual landowners and to the taxpayers of the United States. This year, we assisted in the acquisition of property needed for new or expanded federal courthouses in Buffalo, New York, and Salt Lake City, Utah, and office space in Arlington, Virginia. Through settlements and trials, the Division saved about $32.7 million. We also worked with agencies to avoid the expense of litigation where possible.

Working with the United States Congress on Environmental and Natural Resources Legislation and Related Matters

Through the Department's Office of Legislative Affairs, the Division responds to relevant legislative proposals and Congressional requests, prepares for appearances of Division witnesses before Congressional committees, and drafts legislative proposals in connection with its work, including those implementing litigation settlements. In Fiscal Year 2008, the Division led the Department's review of legislation revising the Lacey Act. Until it was amended this year, the Lacey Act, the country's oldest national wildlife conservation statute, served primarily as an anti-trafficking statute to protect a broad range of fish and wildlife. This year's amendments to the Lacey Act added new enforcement tools with respect to plants, including timber harvested in violation of a foreign country's laws and imported into the United States, as well as products made from illegally harvested plants. The amendments also require importers of plants and plant products to file a declaration upon importation.

The Division has also actively participated in an interagency group led by the Council on Environmental Quality tasked with implementation of the new provisions. Division attorneys have also spoken at conferences and meetings in the United States and internationally to explain the scope of the new Lacey Act provisions to government officials, industry representatives, and others.

Enforcing Environmental Laws Through International Capacity Building

Attorneys from the Division speak at conferences in foreign countries and provide training on a variety of subjects pertaining to civil and criminal environmental enforcement. This year, the Division engaged in such capacity building in Guatemala, France, Norway, Austria, the United Kingdom, the Kingdom of Bahrain, Thailand, Laos, Malaysia, Indonesia, China, and South Africa.

Division attorneys organized and served as instructors in workshops for judges, prosecutors, and investigators in Vietnam and Thailand, addressing prosecuting cases to combat trade in wildlife and wildlife parts in partnership with the Association of Southeast Asian Nations Wildlife Enforcement Network. The Division also conducted training for judges and law enforcement officials in Indonesia on prosecuting cases involving trafficking in illegally harvested timber. The Department of State funded this training pursuant to a Memorandum of Understanding Between the United States and Indonesia on Combating Illegal Logging and Associated Trade. The Division also received funding under the Central America-Dominican Republic-United States Environmental Cooperation Agreement to train judges in Central America on handling environmental enforcement cases.

Division attorneys took part in several rounds of negotiations in Peru and Washington, D.C., with officials of the Government of Peru, to review and provide comments on environmental and forestry laws and regulations being enacted by Peru to meet its obligations under the United States-Peru Trade Promotion Agreement.

Protecting the Interests of the United States in Litigation Involving Third Parties

The Division at times participates as amicus curiae in cases in which the United States is not a party to protect the interests of the United States and its component agencies. Such participation may be in district court, in a court of appeals, or in the Supreme Court; we also participate at times in state court proceedings. The Division has filed briefs in a number of such proceedings in the past year. One example is a Fifth Circuit case involving the res judicata effect of a CWA consent decree on parallel citizen litigation, *ECO v. Dallas*. Here, the Division successfully argued that a citizen suit could not continue after the federal consent decree resolved all claims in the case. Such parallel claims are a recurring issue, and are especially prevalent in the largest and most complex enforcement matters. The favorable *ECO* opinion will help shape the case law and protect the settlement authority of the United States.

Promoting National Security and Military Preparedness

Protecting the Navy's Ability to Use Sonar in Training Exercises

The Division represents the Navy in several cases that challenge the Navy's use of mid-frequency active sonar throughout the world and in specific training exercises off the coast of California and Hawaii, as well as its use of low-frequency sonar, a new technology for anti-submarine warfare that is still in the experimental phase. These high-profile cases are critically important to the nation's security and military readiness. In 2008, our efforts have resulted in significant success.

The plaintiffs challenged the Navy's plans to conduct 11 training exercises off the coast of southern California in *NRDC v. Dep't of the Navy*. Over a six-month period, this litigation involved three preliminary injunction hearings in district court; extensive use of classified materials, including submission of declarations from the Navy's highest ranking operational office (the Chief of Naval Operations); a site visit by the district court judge; the filing of emergency motions to stay before both the district and appellate courts, including one within hours of a court order; two oral arguments before the Ninth Circuit; settlement negotiations before the Ninth Circuit mediator; invocation of a statutory exemption by the President; and the filing of a petition for certiorari before the Supreme Court. Importantly, at each stage of this litigation, the Division has obtained relief allowing the Navy to proceed with training.

In the worldwide Navy sonar case, Division trial attorneys aggressively engaged in jurisdictional discovery initiated by the plaintiffs. Following receipt of a discovery order in July of this year authorizing the Navy to take certain depositions and instructing the plaintiffs to respond to written discovery, the plaintiffs re-started settlement discussions, and a settlement very favorable to the Navy was reached.

Supporting the Strategic Border Initiative

To enhance domestic security, the Illegal Immigration Act requires the Department of Homeland Security (DHS) to build fencing along the United States/Mexico border. The Act requires DHS to designate priority areas where fencing would be most practical and effective in deterring smugglers and aliens attempting to gain illegal entry into the United States and to complete construction of fencing in those areas by December 31, 2008. DHS has identified over 350 miles of priority areas, and the Division has worked closely with DHS and the United States Army Corps of Engineers to facilitate land acquisitions necessary for construction in those areas. This effort has required acquisition by eminent domain of over 300 parcels of land located in Texas, New Mexico, Arizona, and California.

Some of these actions have been appealed to the Fifth Circuit. In *United States v. Muniz & Rivas*, various landowners challenged possession orders issued by the district court granting the United States possession of a temporary right of entry as to the subject properties. Agreeing with the United States, the Fifth Circuit dismissed for lack of appellate jurisdiction.

Acquiring Property to Improve Military Readiness and National Security

The Division also exercised the federal government's power of eminent domain this year to acquire land for military readiness and national security purposes, including for such diverse military installations as the U.S. Southern Command headquarters in Florida; Luke Air Force Base in Arizona; the Department of the Army's National Training Center in Fort Irwin, California; a Special Operations Force riverine training range in Mississippi; Dobbins Air Reserve Base in Georgia; and Travis Air Force Base's Anti-Terrorism Protection project requirements in California.

We also supported the military in other crucial ways, such as reviewing and approving title for the acquisition of a hundred acres of land in Florida and Pennsylvania by the Department of Veterans Affairs for medical facilities and national cemeteries.

The Division also asserted the federal government's power of eminent domain to assess environmental damage and then clean up contaminants stemming from military facilities in Montgomery County, Alabama, and Tooele Army Depot in Utah.

Defending the Army's Chemical Weapons Demilitarization Program

The Division has successfully defended the Army against challenges to its program to destroy aging stockpiles of chemical weapons pursuant to international treaty obligations. In *Sierra Club v. Army*, the plaintiffs challenged the Army's destruction of a chemical nerve agent under RCRA. The destruction process involves neutralizing the deadly liquid agent at one location, then shipping the resulting product to a commercial hazardous waste incinerator. The plaintiffs alleged that the chemical agent is not fully neutralized in the treatment process and that trucking the resulting product thus presents risks. The court ruled in favor of the Army on all counts, allowing destruction of chemical weapons at the facility, which is vital to national security, to be completed without interruption.

An Oregon court issued a largely favorable decision in *G.A.S.P. v. Army*, upholding state-issued permits for the incineration of chemical weapons at the Army's facility in Umatilla, Oregon. The court remanded two relatively minor issues to the state permitting agency, but held that the facility may continue incinerating chemical weapons during the remand because petitioners had not shown that the operations had an adverse effect on public health or the environment.

Defending FBI Weapons Training

In *Pollack v. United States Dep't of Justice*, the plaintiffs sought injunctive relief, $35.2 million for environmental investigation and remediation, and $20 million in tort damages arising out of FBI and other federal agency firearms training at a regional training facility in Illinois, and live-fire training exercises by the Coast Guard in Lake Michigan. They contended that these activities violated the CWA, RCRA, and CERCLA, and constituted a public nuisance under the Federal Tort Claims Act. The court dismissed all claims for lack of subject matter jurisdiction, finding that the plaintiffs lacked individual and organizational standing insofar as they failed to demonstrate individualized harm from the alleged impacts to drinking water drawn from Lake Michigan and to their aesthetic interests in the vicinity of the training facility and Lake Michigan.

PROTECTING INDIAN RESOURCES AND RESOLVING INDIAN ISSUES

Defending Tribal and Federal Interests in Water Adjudications

In Fiscal Year 2008, the Division had continued success in representing the interests of Indian tribes in complex water rights adjudications. The Division was instrumental in reaching a major water rights settlement between the United States, the Soboba Band of Luinseno Indians, and three California water districts. The agreement, approved by Congress on July 23, 2008, and signed into law on July 31, 2008, brought to an end almost 60 years of litigation and over 10 years of settlement negotiations. The settlement provides that the Soboba Band has the paramount right to pump a stipulated amount of water for any use on the reservation, an amount guaranteed by the water districts should interference deplete the Band's groundwater reserves. In addition, the non-Indian water districts have contributed land to the Band and one of the districts will deliver water for the next 30 years to recharge the groundwater.

The Division also successfully defended a comprehensive settlement in *United States v. Washington Dep't of Ecology (Lummi Tribe)*, a lawsuit that sought to determine the Lummi Nation's rights to the use of groundwater underlying the Lummi Peninsula of the Lummi Indian Reservation. The Division, working with the State of Washington and private water users, entered into a settlement which protects the groundwater beneath the Lummi Peninsula.

Protecting Tribal Hunting, Fishing, and Gathering Rights

The Division litigates to protect treaty-based tribal hunting, fishing, and gathering rights. Many years ago, the United States prevailed in establishing the treaty fishing rights of four Columbia River Basin tribes. The taking of those fish, however, impacts anadromous species (those which are born in fresh water, migrate to the ocean to grow into adults, and return to fresh water to spawn) that are listed pursuant to the ESA. In *United States v. Oregon*, the parties, after a decade of negotiations, concluded the Management Agreement for Fish Harvests on the Columbia and Snake Rivers in Washington, Oregon, and Idaho for 2008-2017, which complies with the ESA. This plan will improve fish habitat and allow the tribes to increase their catch as the populations of threatened species increase.

Upholding Agencies' Authority to Implement Indian Policies

The Division continued to achieve considerable success this year in defending the Secretary of the Interior's land trust acquisition authority against numerous constitutional, statutory, and administrative law challenges. The Division also successfully defended decisions by the National Indian Gaming Commission to approve tribal ordinances under the Indian Gaming Regulatory Act. For example, in *City of Vancouver v. Hogen*, the Division obtained dismissal of a lawsuit that challenged the Commission's approval of a Cowlitz Indian Tribe ordinance which addresses important environmental and health/safety issues. In *County of Sauk v. Kempthorne*, the court found that the Secretary acted within statutory authority in approving the contested trust acquisition on behalf of a Wisconsin tribe, which will provide important economic and cultural benefits to the tribe. Finally, in *Michigan Gambling Opposition v. Kempthorne ("MichGO")*, the Division secured rulings from district and appellate courts that rejected environmental and constitutional claims related to a proposed acquisition in Michigan.

Defending Tribal Trust Claims

The Division represents the United States in nearly 100 cases brought by Indian tribes demanding accountings and damages, and alleging breach of trust and other claims relating to funds and non-monetary assets (such as timber rights, oil and gas rights, grazing, mining, and other interests) on some 45 million acres of land. Many of these cases are in settlement negotiations and others are in the early stages of pre-trial preparation. The cases are complex and cover many decades of economic activity on tribal reservations. The Division has enjoyed great success in a program of engagement with the tribes on their claims and has fairly balanced its duties to defend client programs with an obligation to make whole any tribes wronged by asset management practices. The Division has settled some cases, had others dismissed on procedural grounds, and is prepared to go to trial in yet others.

Litigating under the Indian Gaming Laws

With the Indian gaming revenue at tribal casinos exceeding the combined revenue of Nevada and Atlantic City casinos, the Division continues to be involved in litigation defending the Secretary's actions related to Indian gaming. The Division was successful in securing a dismissal of *State of Florida and Alabama v. Kempthorne*, which challenged the constitutionality of regulations implementing federal Indian gaming laws. In three separate actions, we successfully opposed emergency injunctive relief seeking to prevent a tribal-state compact from going into effect in the State of Florida. In doing so, we secured very favorable language that gives the Interior Secretary discretion and flexibility in the Department's sensitive role in approving tribal-state compacts.

Navigating the Tribal Acknowledgment Process

No role of the Bureau of Indian Affairs in the Department of the Interior is more sensitive and at times more controversial than its responsibility to make decisions on whether a group should be federally recognized as a tribe. The decision means the group becomes sovereign over its members and internal disputes. A federally recognized tribe gains distinct federal benefits and may no longer be subject to state civil regulatory laws on its lands. Federal recognition also allows the tribe to establish Indian gaming activities that are consistent with federal law. As a result, the acknowledgment process has become a magnet for litigation. The Division had two prominent successes this year in upholding the acknowledgment process. In the first case, *MOWA Band of Choctaw Indians v. United States*, the court upheld the Interior Department's administrative consideration of the plaintiff's application for recognition and found that the Department had effectively notified the plaintiff of its decision, thus commencing the statute of limitations period which barred plaintiff's action. In the second case, *Schaghticoke Tribal Nation v. Kempthorne*, the plaintiff charged that political interference led to the denial of its petition for recognition. The court noted that, while the evidence showed that political actors had exerted public and private pressure on the Department, none of that political activity actually affected the outcome of the decision on the acknowledgment petition.

SUPPORTING THE DIVISION'S LITIGATORS

Expanding and Upgrading Technology Services and Resources

High priority was placed on technology services and resources in 2008. In keeping with the Division's unique responsibility to lead the Department in areas of environmental and energy efficiency, we replaced aging IT infrastructure and deployed new "virtual server" technology, which allowed the Division to purchase 37% fewer servers. The new servers include an energy-saving technology that exceeds EPA's Energy Star requirements. Together, these efficiencies will reduce the Division's power requirements and heat output by 50%. Technical support for Division staff also received a boost this year when the computer desktop support contract (help desk) was restructured to provide service 24 hours a day, 7 days a week.

The Division acquired software to save attorney time and effort during document review. We also continued to provide government attorneys with assistance in electronic discovery matters, including through training, client agency consultations, and documented standard practices.

Several new features were added to the Division's Internet site, www.usdoj. gov/enrd, including a subscription service by which members of the public can sign up for email alerts about items of interest.

The Division's internal Intranet site contains new connectivity tools to help employees more quickly access information and complete tasks necessary for their jobs. One such tool completed in 2008 is a web-based expert witness and litigation consultant tracking system. The Division also improved the ability of trial teams to collaborate on case materials with agency counsel, investigators, and expert witnesses by routinely making case information available through our secure Internet portal.

We completed the Division's conversion to JUTNet, a managed enterprise local and wide area network that provides secured data telecommunications services to Division sites in Washington, D.C., and all field office locations. This Department-wide system meets new federal information sharing and security requirements.

Greening the Division

The Division is an active participant in the Department's Greening the Government effort. This year, the Division's Greening the Government Committee took concrete steps to reduce the environmental footprint of the Division, and also conducted research that should lead to future reductions in energy use and material waste.

Much of the committee's work to date has focused on informational outreach to Division employees to encourage environmentally-friendly behaviors while at work. "Best Practices" memoranda for paper conservation and energy use were distributed to staff, and weekly messages on the Division's Intranet have provided other tips for being "green" at work. To reduce paper waste, the committee has posted informational placards near recycling bins and the Division will be purchasing paper with a higher percentage of recycled content. And in a step that will conserve paper, all sections of the Division have elected to standardize double-sided printing as their default printing option. The Division has also enrolled in the EPA-ABA Climate Challenge, a program for law offices seeking to document improvements in their environmental performance.

Supporting Litigation

The Office of Litigation Support participated in the Division's most complex cases in 2008, making use of contract staff and in-house expertise. Careful attention to case requirements analysis and attorney trial team needs allowed the Division to apply a $27 million litigation support budget to document management, processing, staffing, training, and trial support services for over 300 matters and initiatives.

As part of ongoing efforts to provide better, more efficient services to attorneys at the most affordable rates for the government, we relocated contract litigation support operations for two of the Division's largest cases to a large, convenient, and less expensive support center location, later integrating other Division cases into the new support center. We closed a large document center in Boise, Idaho, and consolidated all western water case activities in Denver, Colorado.

On-site trial support was provided for major civil and criminal cases across the country, in locations as diverse as Cincinnati, Ohio, Los Angeles, California, Portland, Oregon, Providence, Rhode Island, and Corpus Christi, Texas.

In: Enforcing Federal Pollution Control Laws
Editor: Norbert Forgács

ISBN: 978-1-60876-082-4
© 2010 Nova Science Publishers, Inc.

Chapter 2

ENVIRONMENTAL ENFORCEMENT: EPA NEEDS TO IMPROVE THE ACCURACY AND TRANSPARENCY OF MEASURES USED TO REPORT ON PROGRAM EFFECTIVENESS[*]

United States Government Accountability Office

September 18, 2008

The Honorable John Dingell
Chairman
Committee on Energy and Commerce

The Honorable Bart Stupak
Chairman
Subcommittee on Oversight and Investigations Committee on Energy and Commerce
House of Representatives

Subject: *Environmental Enforcement: EPA Needs to Improve the Accuracy and Transparency of Measures Used to Report on Program Effectiveness*

As part of its mission to protect human health and the environment, the Environmental Protection Agency's (EPA) enforcement office maintains civil and criminal enforcement programs to help enforce the requirements of major federal environmental laws such as the Clean Air Act and the Clean Water Act. EPA's civil and criminal enforcement programs work with the Department of Justice (DOJ), and in some cases states, to take legal actions to bring polluters into compliance with federal laws. While civil enforcement actions require polluters to pay penalties and take other corrective actions, criminal enforcement actions also

---

[*] This is an edited, reformatted and augmented version of a U. S. Government Accountability Office publication, Report GAO-08-111R, dated September 2008.

may include imprisonment. EPA's enforcement office sets national priorities to focus resources on significant environmental risks and non-compliance patterns; prepares nationally significant civil and criminal cases for legal action by DOJ; uses 10 regional offices to implement civil enforcement actions on a day-to-day basis; and pursues criminal violations of environmental laws through its criminal enforcement office. The agency exercises its authority to independently pursue some violators through administrative proceedings—civil administrative actions—and to refer significant matters to DOJ when it believes cases need to be filed in federal court as civil judicial actions.[1] DOJ is responsible for prosecuting and settling civil judicial and criminal enforcement cases.

EPA relies on a variety of measures to assess and report on the effectiveness of its civil and criminal enforcement programs. For example, EPA relies on assessed penalties that result from enforcement efforts among its long-standing measurable accomplishments. The agency uses its discretion to estimate the appropriate penalty amount based on individual case circumstances. EPA has developed penalty policies as guidance for determining appropriate penalties in civil administrative cases and referring civil judicial cases. The policies are based on environmental statutes and have an important goal of deterring potential polluters from violating environmental laws and regulations. The purpose of EPA's penalties is to eliminate the economic benefit a violator gained from noncompliance and to reflect the gravity of the alleged harm to the environment or public health.[2]

Like other federal agencies, EPA has established results-oriented goals and performance measures. Two of the major performance measures for civil enforcement, according to EPA, are (1) the value of injunctive relief—the monetary value of future investments necessary for an alleged violator to come into compliance, and (2) pollution reduction—the pounds of pollution to be reduced, treated, or eliminated as a result of an enforcement action.[3] EPA told us these two measures, as well as penalties, should be considered when assessing the overall impact of its enforcement actions. EPA relies on these measures, among others, in pursuing its national enforcement priorities and overall strategy of fewer, but higher impact, cases. Unless these measures are meaningful, Congress and the public will not be able to determine the effectiveness of the programs. Therefore, it is important to understand how they are determined and the extent to which they accurately reflect EPA's accomplishments.

In this context, we agreed to report on (1) amounts of civil and criminal penalties assessed in recent years and how EPA calculates and reports on these outcomes, (2) the value of injunctive relief and amounts of pollution reduction and how EPA calculates and reports on these outcomes, and (3) factors that influence EPA's process in achieving enforcement outcomes. This chapter recommends steps that EPA should take to improve the transparency and accuracy of its reports to Congress and the public when reporting on the effectiveness of its enforcement programs.

In conducting our work, we reviewed agency documents such as guidance and policy statements as well as reports to Congress and the public. In addition, we reviewed EPA information associated with the case that the agency identified as resulting in the largest value of injunctive relief in its history. We also met with EPA headquarters and regional officials, DOJ officials, and non-profit groups concerned with environmental enforcement. We reviewed EPA reports of monetary accomplishments presented in nominal dollars and adjusted these amounts for inflation when determining the extent of trends in the data through statistical analysis. We primarily focused our penalty analysis on fiscal years 1998 though 2007 since EPA officials said they were confident in the data within most of the period and in our judgment the most recent 10-year period appeared to be a reasonable time frame. Further, we were able to perform some analysis of the data reliability for most of those years by comparing amounts in EPA's database available only to government officials and amounts reported to the public. We conducted this performance audit in accordance with generally accepted government auditing standards from January 2008 through September 2008. Those standards require that we plan and perform the audit to obtain sufficient, appropriate evidence to provide a reasonable basis for our findings and conclusions based on our audit objectives. We believe that the evidence obtained provides a reasonable basis for our findings and conclusions based on our audit objectives.

RESULTS IN BRIEF

Total penalties assessed by EPA, when adjusted for inflation, declined from $240.6 million to $137.7 million between fiscal years 1998 and 2007. We identified three shortcomings in how EPA calculates and reports penalty information to Congress and the public. Specifically, EPA is:

- Overstating the impact of the enforcement programs by reporting penalties assessed against violators rather than actual penalties received by the U.S. Treasury.
- Reducing the precision of trend analyses by reporting nominal rather than inflation-adjusted penalties, thereby understating past accomplishments.
- Understating the influence of its enforcement programs by excluding the portion of penalties awarded to states in federal cases.

In contrast to penalties, we found that both the value of estimated injunctive relief and the amount of pollution reduction reported by EPA generally increased. The estimated value of injunctive relief increased from $4.4 billion in fiscal year 1999 to $10.9 billion in fiscal year 2007, in 2008 dollars. In addition, estimated pollution reduction commitments amounted to 714 million pounds in fiscal year 2000 and increased to 890 million pounds in fiscal year 2007. However, we identified several shortcomings in how EPA calculates and reports this information. We found that generally EPA's reports do not clearly disclose the following:

- Annual amounts of injunctive relief and pollution reduction have not yet been achieved. They are based on estimates of relief and reductions to be realized when violators come into compliance.
- Estimates of the value of injunctive relief are based on case-by-case analyses by EPA's technical experts, and in some cases the estimates include information provided by the alleged violator.
- Pollution reduction estimates are understated because the agency calculates pollution reduction for only 1 year at the anticipated time of full compliance, though reductions may occur for many years into the future.

Finally, we identified factors that affect EPA's process in achieving penalties, injunctive relief, and pollution reduction. For example, DOJ, not EPA, is primarily responsible for prosecuting and settling civil judicial and criminal enforcement cases. Therefore, EPA does not have ultimate control of enforcement outcomes.

We are recommending that the EPA Administrator take a number of actions to disclose more information when reporting penalties and estimates of the value of injunctive relief and pollution reduction.

While Assessed Penalties Declined between Fiscal Years 1998 and 2007, There Are Three Shortcomings in How EPA Calculates and Reports Penalties

From fiscal years 1998 to 2007 total inflation-adjusted penalties declined when excluding major default judgments,[4] and we identified three shortcomings in how EPA calculates and reports on these outcomes. Total penalties reported by EPA are the sum of assessed penalties resulting from EPA's civil administrative, civil judicial, and criminal enforcement actions. When adjusted for inflation, total assessed penalties were approximately $240.6 million in fiscal year 1998 and $137.7 million in 2007. Civil judicial penalties are the largest source of assessed penalties, accounting for about 45 percent of the total (see table 1).

While these total inflation-adjusted penalties tended to decline during this period, the trend exhibits only marginal statistical significance.[5] The data, according to EPA, include penalties for three major cases totaling $227.2 million in 2008 dollars that EPA does not expect the federal government to collect due to default judgments, which represent uncontested cases where courts awarded the statutory maximum penalty requested by EPA and DOJ. Figure 1 highlights the three penalties in fiscal years 2004 through 2006, ranging from $33.8 million to $104.4 million in 2008 dollars, and illustrates the trend for this period.

Table 1. Assessed Penalties Reported by EPA, Adjusted for Inflation

Constant 2008 dollars in millions				
Fiscal Year	Civil Judicial	Administrative	Criminal	Total
1998	$82.9	$36.6	$121.1	$240.6
1999	180.8	32.7	78.8	292.3
2000	68.1	36.3	151.3	255.7
2001	122.2	28.6	113.9	264.7
2002	75.6	30.6	73.7	180.0
2003	83.6	28.2	82.2	194.0
2004	137.1	31.3	53.1	221.5
2005	139.3	29.3	109.5	278.1
2006	86.4	44.4	45.4	176.2
2007	41.0	31.7	65.0	137.7
Total	$1,017.0	$329.6	$894.1	$2,240.7
Percent of total	45.4%	14.7%	39.9%	100.0%

Source: GAO analysis based on EPA data.
Note: Numbers may not add due to rounding.

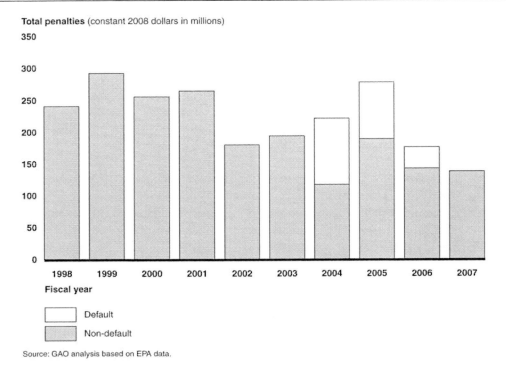

Figure 1. Total Inflation-Adjusted Assessed Penalties, Fiscal Years 1998 through 2007, Default Cases Identified.

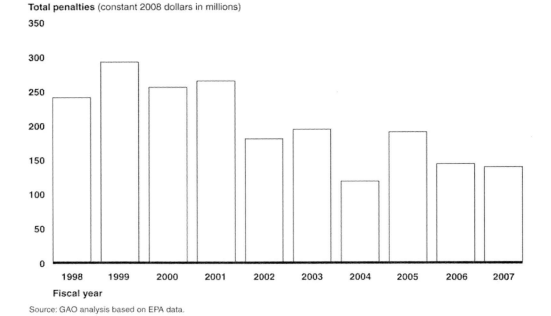

Figure 2. Total Inflation-Adjusted Assessed Penalties, Fiscal Years 1998 through 2007, Less Default Cases.

When excluding these default judgments, total inflation-adjusted penalties exhibit a statistically significant downward trend between fiscal years 1998 and 2007 (see figure 2).

Excluding certain years or choosing different timeframes for analysis could remove the appearance of a downward trend. While our analysis focused on fiscal years 1998 to 2007, when reviewing EPA's reported data since 1974, we recognized that total penalties increased until the late 1990s and stopped rising thereafter (see enclosure I).

We identified three problems in how EPA calculates and reports penalties that may inhibit the accuracy and transparency of EPA's reporting:

- EPA does not report the actual amounts of penalties received by the U.S. Treasury. This may overstate the impact of the enforcement programs by reflecting penalties that have not, or will not, be collected. For example, EPA identified three major civil judicial cases in recent years that generated significant amounts of assessed penalties through default judgments. These penalties are unlikely to ever be collected, and the removal of these penalties results in a significant reduction in the overall level of penalties reported by EPA.
- When reporting penalties over time, EPA presents nominal amounts that are not adjusted for inflation and, therefore, understate past accomplishments. According to OMB, economic analyses are often most readily accomplished using real or constant-dollar values to measure benefits and costs in units of stable purchasing power. Therefore, to evaluate real trends in penalties, it is necessary to remove the effect of price changes in the reported nominal penalties by adjusting for inflation.
- The penalty amounts EPA reports do not include portions of penalties awarded to states in federal cases in which states also participated. EPA indicated that states also participate in many federally-led enforcement cases that result in penalties paid to both EPA and the states. However, EPA reports only the amount of penalties assessed for payment to the federal government, thereby understating the effects of its enforcement efforts on defendants. For example, in 1999 EPA and the State of California jointly settled an enforcement case with a major commercial diesel engine manufacturer for alleged violations of the Clean Air Act. The company agreed to pay a total of $25 million in penalties, of which $18,750,000 was paid to the federal government and $6,250,000 was paid to the State of California. However, only EPA's share of $18,750,000 was included in its reporting of penalties.

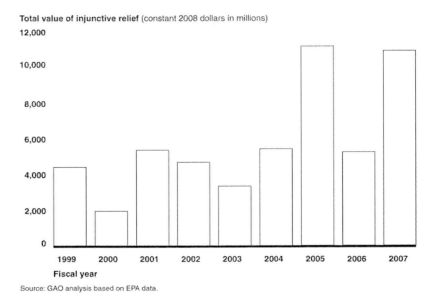

Figure 3. Total Inflation-Adjusted Value of Estimated Injunctive Relief, Fiscal Years 1999 through 2007

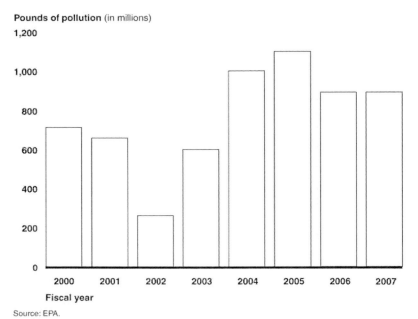

Note: The data from fiscal years 2000 to 2002 are based on EPA's *Performance and Accountability Report for Fiscal Year 2007*, however EPA's enforcement office reports the data from fiscal years 2003 to 2007, citing improved data quality assurance starting in fiscal year 2003. We cannot be certain of the extent that the revised methodology affected the reported levels of pollution reduction after 2003 compared to prior years.

Figure 4. Total Estimated Pounds of Pollution To Be Reduced or Treated, Fiscal Years 2000 through 2007

40 United States Government Accountability Office

Shortcomings in How EPA Reports Measures of Injunctive Relief and Pollution Reduction May Inhibit Accuracy and Transparency of Reporting

The value of estimated injunctive relief, when adjusted for inflation, has increased from $4.4 billion in fiscal year 1999—the earliest period for which EPA has reported the measure—to $10.9 billion in fiscal year 2007 (see figure 3).

Estimated pollutant reduction commitments amounted to 714 million pounds in fiscal year 2000, peaked at 1.1 billion pounds in fiscal year 2005 and decreased to 890 million pounds in fiscal years 2006 and 2007 (see figure 4).

In reviewing the value of injunctive relief and pollution reduction amounts reported by EPA, we identified several shortcomings in how EPA calculates and reports these outcomes that may inhibit the accuracy and transparency of EPA's reporting. The following shortcomings are manifested in EPA reports to Congress and the public, such as (1) annual accomplishments reports on enforcement performance and accountability, and (2) reports comparing EPA's goals and accomplishments under the Government Performance and Results Act:[6]

- EPA calculates estimated rather than actual amounts of pollution reduction based on a 1-year period in the future at the anticipated time of full compliance, and the value of injunctive relief based on the monetary value of an alleged violator's estimated future investments to come into compliance. However, the agency's reports do not always make it clear that these amounts have not been achieved. For example, EPA's fiscal year 2007 accomplishment report on enforcement referred to the largest civil enforcement actions for just three priority areas alone that "...achieved more than 400 million pounds of pollutant reductions and more than $7 billion in injunctive relief and supplemental environmental projects."[7] However, for the most part, those amounts were estimates of future anticipated results, such as an estimated defendant's future costs over several years, and do not represent actual accomplishments. Similarly, EPA's annual performance and accountability report, referring to total pollution reduction, states "EPA has reduced, treated or eliminated 890 million pounds of pollution through enforcement actions in fiscal year 2007." However, not all of those pollution reductions actually occurred in 2007.
- EPA does not disclose in its estimates of the value of injunctive relief how the estimates are derived. In estimating the value of injunctive relief, EPA technical staff rely on their professional judgment without any agency guidance or systematic processes, and in cases where they are available they rely on estimates of alleged violators. For example, in one major settlement EPA estimated that the value of injunctive relief would total $4.6 billion, the largest injunctive relief amount in the agency's history. The purpose of the injunctive relief in this case is to reduce future air pollutants from several coal-fired generating plants of a power company.[8] EPA officials told us they based the estimated value on advice from their technical experts and examination primarily of a 3-page document[9] the company provided through discovery.[10] Furthermore, EPA officials said

defendants are not always compelled to provide information that the agency could use to estimate future costs of compliance.

- EPA's estimates of pollution reduction may be understated because EPA reports only 1- year of estimated pollution reduction at the anticipated time of full compliance for a given case, although reductions may occur for many years into the future. In addition, EPA's estimates do not account for incremental reductions in the years leading up to full compliance.
- The estimated pounds of pollution reduced, treated, or eliminated does not reflect the varying toxicity of the types of pollution represented by the measure. For example, EPA officials said that the amount of mercury to be reduced in the atmosphere as a result of enforcement efforts may be a small number of pounds when compared to other pollutants, but mercury is a more toxic substance than many other pollutants that are included in the measure. EPA officials said they recognize this issue and they are working to address it.

Other Factors Influence EPA's Process for Achieving Enforcement Outcomes

EPA's process for achieving annual results in terms of penalties, estimated value of injunctive relief, and amounts of pollution reduction is influenced by many other factors. While the following list is not comprehensive, it describes some of the significant aspects of the legal and policy environment that could affect the outcomes:

- The Department of Justice (DOJ), not EPA, is primarily responsible for prosecuting and settling civil judicial and criminal enforcement cases. The Attorney General is charged by statute with conducting and supervising litigation to which the United States, or its departments or agencies, is a party, including cases referred by EPA.11 Once cases are referred, EPA officials stated that they continue to participate in all civil and many criminal cases. For each case, DOJ must weigh the litigation risks that affect the likely outcome at trial in making its decisions on whether or how to settle. Consequently, DOJ officials said EPA's proposed penalty estimates do not govern DOJ's decisions. DOJ, like EPA, considers applying penalties as described in the relevant environmental statutes. EPA and DOJ officials say they cooperate and reach mutually agreeable decisions on civil judicial cases. For example, DOJ officials said both agencies sign the settlement agreements. However, EPA does not have ultimate control over the enforcement outcomes.
- Executive Order 12988 directs DOJ, whenever feasible, to seek settlements before pursuing civil judicial actions against alleged violators. According to DOJ officials, the Executive Order encourages negotiations prior to the onset of litigation and, thereby, improves the ability of the United States to achieve favorable enforcement outcomes.
- Unclear legal standards, as illustrated in the following examples, have hindered EPA's enforcement efforts. Agency officials told us a 2006 Supreme Court

decision, Rapanos v. United States, generally made it more difficult for EPA to take enforcement actions because the legal standards for determining what is a "water of the United States" were not clear. This uncertainty required EPA to gather significantly more evidence to establish Clean Water Act jurisdiction in those cases where alleged violators discharged to waters of the United States. In a March 2008 memorandum, EPA's Assistant Administrator for Enforcement and Compliance Assurance said the Court decision and EPA's resulting guidance "negatively affected approximately 500 enforcement cases." For example, the official said EPA's regions decided not to pursue formal enforcement in about 300 instances where there were potential violations because of jurisdictional uncertainty.

- A rule change can affect the process for achieving enforcement outcomes. For example, according to an EPA Office of the Inspector General (OIG) report in 2004, a New Source Review rule change finalized in October 2003 "seriously hampered (EPA) settlement activities, existing enforcement cases, and the development of future cases" due largely to EPA's revised definition of routine maintenance.12 Under the revised rule the definition of routine maintenance allowed utilities to undertake projects representing a greater percentage of the cost of replacing a power unit—up to 20 percent—without being subject to the New Source Review requirements. According to the OIG, while EPA officials said the rule change was not retroactive, the change was so dramatic, that even though a court in December 2003 issued a stay delaying implementation of the rule, EPA's underlying legal arguments may have been weakened.13 For example, three utilities said enforcement under a court-imposed remedy should be heavily reduced because their actions would not be a violation under the new rule. Furthermore, at the time the IG report was issued in September 2004, no new enforcement actions had been taken against coal- fired utilities alleged to have violated the old rule because of the new rule's impact on EPA's leverage in settlements or court remedies, according to the OIG. The decline in cases between 2002 through 2003 is also, according to EPA, due to the agency not initiating coal-fired power plant cases during the proposal and promulgation of the new rule. EPA officials said they initiated or concluded eight cases under the old rule since 2003.

CONCLUSIONS

Pursuing administrative, civil, or criminal action against a suspected polluter is a complex undertaking that often lasts years. While EPA's reported outcomes of enforcement efforts help inform Congress, the public, and EPA management about EPA's progress in prosecuting those who violate federal environmental laws, certain aspects of how EPA reports the data may undermine the transparency and accuracy of its reported outcomes and cause EPA to both over and under-report its enforcement achievements. Taken as a whole, these various shortcomings hamper the transparency and accuracy of EPA's reporting and create the potential for Congress and the public to misunderstand the agency's enforcement outcomes.

RECOMMENDATIONS FOR EXECUTIVE ACTION

To improve the transparency and accuracy of its reports to Congress and the public when reporting on the effectiveness of the enforcement programs, we recommend that the EPA Administrator take the following six actions:

- When reporting the amount and nature of penalties stemming from enforcement actions, disclose (1) penalties in a manner that clearly indicates that they are assessed rather than collected penalties, (2) penalties collected as well as assessed by the federal government, (3) time series data that are adjusted for inflation, and (4) states' share of penalties in federal cases.
- When reporting other major outcome measures of civil enforcement efforts, clearly disclose (1) that the monetary value of injunctive relief is based on estimates of future amounts that defendants expect to spend to achieve outcomes, as agreed in consent decrees, and (2) that the pounds of pollution reduced represent the anticipated reduction for a 1-year period at the anticipated time of compliance.

David C. Maurer
Acting Director, Natural
Resources and Environment

ENCLOSURES I

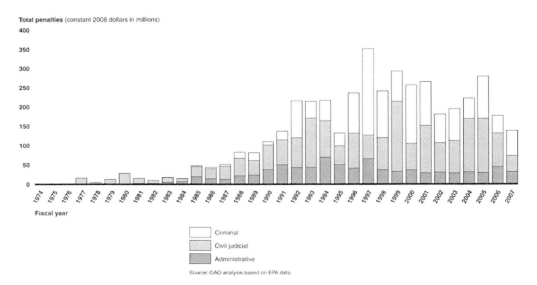

Figure 5. Total Inflation-Adjusted Assessed Penalties, Fiscal Years 1974 through 2007, by Type

ENCLOSURES II

Comments from the Environmental Protection Agency

UNITED STATES ENVIRONMENTAL PROTECTION AGENCY
WASHINGTON, D.C. 20460

SEP 11 2008

ASSISTANT ADMINISTRATOR
FOR ENFORCEMENT AND
COMPLIANCE ASSURANCE

Mr. David C. Maurer
Acting Director, Natural Resources and Environment
U. S. Government Accountability Office
Washington, DC 20548

Dear Mr. Maurer:

Thank you for the opportunity to comment on "Environmental Enforcement: EPA Needs to Improve the Accuracy and Transparency of Measures Used to Report on Program Effectiveness," Project Number GAO-08-1111R. The Office of Enforcement and Compliance Assurance (OECA) appreciates the work of GAO in preparing this report and generally accepts the recommendations provided. OECA is proud of its accomplishments in protecting public health and the environment, and agrees that clarity and transparency in the reporting of our results is important. We appreciate GAO's suggestions for improving clarity and transparency.

In our response below we address the specific recommendations and provide some additional substantive comments.

This draft report evaluates the accuracy and transparency of the performance measures OECA uses to report on program effectiveness. The performance measures examined include penalties, value of injunctive relief, and pounds of pollution estimated to be reduced, treated or eliminated. Below is OECA's response to recommendations and suggested corrections to technical inaccuracies.

I.   OECA Response to Draft Recommendations

Recommendation 1: Clearly indicate in public reports and press releases that penalties reported are assessed.

GAO is correct that EPA reports penalties assessed rather than penalties collected. The purpose of reporting penalties assessed is to communicate to the public the consequences of noncompliance and to create a general deterrent effect that helps us achieve our mission.

Environmental Enforcement

Response: OECA will monitor its press releases, Annual Results reports and other public documents to ensure that it is clear that penalties reported for a particular case or year are penalties assessed.

Recommendation 2: Report penalties collected as well as assessed.

OECA continues to believe that reporting penalties assessed is the key measure for reporting to the public. While we agree that penalties collected is a useful internal management measure, we do not believe that penalties collected should be publicly reported when EPA announces individual case settlements or in its Annual Results. As a practical matter, amounts collected are not actually known for some time after the settlement is announced. Administrative penalties are collected by EPA's Office of the Chief Financial Officer (OCFO). Civil judicial and criminal penalties are collected by the Department of Justice (DOJ) through the individual U.S. Attorneys' offices. The amounts collected are tracked by these many different offices. We have begun making changes to our information systems and operating procedures that will enable us to track when judicial penalties have been paid in full under judicial Consent Decrees, and expect to begin collecting this information in FY09.

Response: OECA continues to regard the reporting of assessed penalties alone to be of greater deterrent value than reporting both assessed and collected penalties. However, we will discuss this recommendation with EPA's Office of Chief Financial Officer and the Department of Justice.

Recommendation 3: Provide time-series data adjusted for inflation.

Response: OECA concurs and will begin reporting this information for FY 2008.

Recommendation 4: Report states' share of penalties in federal cases.

The report is correct that EPA has not included state penalties in reporting of annual results. This conservative approach was taken to avoid claims that we overstated our results. If we had included the state share of penalties in federal cases, our penalty numbers would have been higher during the period reviewed by GAO. However, OECA recognizes that the State penalty amounts we obtain in our settlements do contribute to deterrence.

Response: Starting in FY 2009, OECA will report states' share of penalties assessed. This information will be reported separately from federal penalty amounts assessed.

United States Government Accountability Office

Recommendation 5: Make clear that the value of injunctive relief reported is based on estimates of future amounts that defendants expect to spend to achieve outcomes as agreed in Consent Decrees.

OECA strives to ensure that it is clear that its injunctive relief values reflect estimates of future commitments by the defendants to achieve compliance as specified in the consent decree.

Response: OECA will redouble its efforts to ensure that our reports make clear what this measure represents.

Recommendation 6: Clearly disclose that the pounds of pollution reduced represent the anticipated reduction for a one-year period at the anticipated time of completion.

The pollutant reductions in many cases can be expected to continue for many years or indefinitely. This poses a question of what future time period OECA should use in projecting and reporting the pollutant reduction results. OECA has chosen to limit its projections to the one year period following completion of the injunctive relief. OECA has adopted this approach to avoid the potential for overstating results. OECA has endeavored to make this clear in its reporting of results but acknowledges that a full explanation has not been present in all documents.

Response: OECA agrees that its report of results should make clear the time period over which estimated results are projected.

II. Response to Technical Inaccuracies

Page 1 of draft report, footnote 1: "EPA generally depends on DOJ…in some federal lawsuits."

Response: Revise footnote 1 to read, "Congress has limited EPA's authority to pursue violations in an administrative forum under some of the environmental statutes that EPA is responsible for enforcing. For instance, under the Clean Air Act, EPA may pursue penalties in an administrative forum only if the total penalty sought does not exceed $270,000 (as adjusted by the Civil Monetary Penalty Inflation Adjustment Rule) and the first alleged date of violation occurred no more than 12 months prior to the initiation of the administrative action, except where EPA and the Department of Justice "jointly determine that a matter involving a larger penalty amount or longer period of violation is appropriate for administrative penalty action." See 42 U.S.C. § 7413(d)(1). Additionally, Congress limited EPA's compliance order authority under the Clean Air Act in the administrative forum. EPA may only issue an order "to require the person to whom it was issued to comply with the requirement as expeditiously as practicable, but in no event longer than one year after the date the order was issued, and shall be nonrenewable." See 42 U.S.C. § 7413(a)(4). Where a matter does not satisfy the above statutory criteria, it is not appropriate for the administrative forum, EPA will have to refer the matter to the Department of Justice for enforcement in the civil judicial forum."

Page 4, second paragraph, last sentence: "Specifically, we found that... do not clearly...":

Response: Revise to read: "Specifically we found that EPA's reports do not in every instance clearly disclose the following:"

Page 5, first paragraph, second sentence: "...DOJ, not EPA, is responsible for prosecuting and settling civil judicial and criminal enforcement cases. Therefore, EPA does not have ultimate control of enforcement outcomes."

Response: Revise to read, "...DOJ, working with EPA, is primarily responsible for prosecuting and settling civil judicial and criminal enforcement cases and for collection of civil judicial penalties. Therefore, EPA does not have ultimate control of all enforcement outcomes or the collection of all penalties.

Page 10: "...peaked at 1.1 billion pounds in fiscal year 2005 and leveled off at 890 million pounds..."

Response: Delete "leveled off at" and replace with "has been". Given the variance from year to year, and a record increase already achieved for FY 2008 greater than that achieved in FY 2005, FY 2006 and FY 2007 combined, the values for 2006 and 2007 do not reflect a "leveling off."

Page 12, second paragraph, last sentence, "EPA officials told us they based the estimate on advice from their technical experts... through discovery."

Response: Revise to read, "EPA officials told us . . .from their technical experts, examination of a 3-page document the company provided through discovery, and that their decision was further informed by pollution control planning documents obtained during discovery, representations made in the litigation and transcribed statements made to the federal court judge supervising confidential settlement negotiations."

Page 13, bottom paragraph, first, third and fifth sentences: "...DOJ, not EPA, is responsible for prosecuting...cases... Consequently, DOJ officials said EPA's proposed ... do not govern DOJ's decisions. While EPA and DOJ...enforcement outcomes."

Response: As written, this statement implies that EPA doesn't play a role. Revise first sentence, to read, "...DOJ, with participation from EPA on all civil and many criminal matters, is primarily responsible for prosecuting... Revise third sentence to read, "Consequently, DOJ officials said EPA's proposed penalty estimates do not exclusively govern DOJ's decisions. Revise fifth sentence by deleting "While" and "EPA does not have ultimate control over the enforcement outcomes" so that sentence reads, "EPA and DOJ officials say they cooperate and reach mutually agreeable decisions on civil judicial cases."

Page 14, bottom paragraph, last sentence: "Second, no court has considered an award of civil penalties... in such a case."

Response: The last sentence is inaccurate and should be deleted. Two federal courts have imposed penalties in NSR cases. See U.S. v. Louisiana-Pacific Corp., 682 F. Supp. 1141 (D. Colo. 1988); and U.S. v. Chevron U.S.A., Inc., 639 F. Supp. 770 (W.D. Tex. 1985). Second, the sentence goes on to say that "so there is no precedent on how high a penalty a court may award in such a case." Although these two cases are twenty years old, they do supply some evidence as to how courts addressed the amount of civil penalties to award for NSR violations. Finally, the statement does not support the GAO's theory that "unclear legal standards have hindered EPA's enforcement efforts."

Page 15, first paragraph, third sentence: "While EPA... retroactive."

Response: This sentence should be deleted.

Page 15, first paragraph:

Response: Add the following sentence before the last sentence in the paragraph: "The decline in cases between 2002 – 2003 is also, according to EPA, due to EPA not initiating coal-fired power plant cases during the proposal and promulgation of the new rule."

Page 15, last sentence: "EPA officials said they concluded a number... since 2004."

Response: Revise to read, "EPA officials said they initiated and/or concluded ... since 2003." See case list provided below on page 6.

Environmental Enforcement

CAA NSR Power Plant Cases Initiated (Either Filed or Settled)
2003-2008

Case Name	Date Filed or Settled
Virginia Electric Power Company	4/17/03
Wisconsin Electric Power Company	4/27/03
South Carolina Public Service Authority (Santee Cooper)	3/16/04
Minnkota Power Cooperative	4/24/06
Kentucky Utilities	3/12/07
Nevada Power	6/13/07
East Kentucky Power Cooperative	7/2/2007
Salt River Project	8/12/2008

If you have any questions concerning our response please contact me at 202/564-2440 or Margaret Schneider, Director of Administration and Policy, at 202/564 2530.

Sincerely,

Granta Y. Nakayama

GAO's MISSION

The Government Accountability Office, the audit, evaluation, and investigative arm of Congress, exists to support Congress in meeting its constitutional responsibilities and to help improve the performance and accountability of the federal government for the American people. GAO examines the use of public funds; evaluates federal programs and policies; and provides analyses, recommendations, and other assistance to help Congress make informed oversight, policy, and funding decisions. GAO's commitment to good government is reflected in its core values of accountability, integrity, and reliability.

End Notes

[1] EPA generally depends on DOJ to file a complaint in court when EPA seeks penalties greater or compliance periods longer than the administrative limits imposed by statutes. For example, under the Clean Air Act, the maximum amount that may be sought in a single administrative enforcement action is $270,000, adjusted for inflation, although higher amounts may be pursued with joint approval of the EPA Administrator and the Attorney General. Also, under the Act, EPA may only issue an administrative compliance order requiring the violator to comply as expeditiously as practicable, but in no event longer than 1 year after the date of the order. In addition, states may also participate as plaintiffs in some federal lawsuits.

[2] Violators frequently obtain an economic benefit by avoiding or delaying necessary compliance costs, by obtaining an illegal profit, by obtaining a competitive advantage, or by a combination of these factors. EPA has

developed an economic model for assisting the agency in determining the portion of a penalty that should be attributable to a polluter's economic benefit from a violation.

[3] The Government Performance and Results Act of 1993 (GPRA) requires that each agency report annually to Congress on the results of its activities in each fiscal year. Program managers use these measures as short-term indicators of program performance and in longer-term trend analyses.

[4] A default judgment is a binding judgment in favor of the plaintiff when the defendant has not responded to a civil complaint.

[5] The tests of statistical significance cited in this paragraph are based on simple linear regression analyses of penalty amounts as a function of year. When analyzing total inflation-adjusted penalties for fiscal years 1998 through 2007, the trend is marginally significant. When default cases are removed for 1998 through 2007, and when total inflation-adjusted penalties are analyzed from 1974 through 2007, the trends are statistically significant at the less than 0.05 level.

[6] EPA, *FY 2007 Office of Enforcement and Compliance Assistance Accomplishments Report* and *Performance and Accountability Report for Fiscal Year 2007.*

[7] As part of a settlement, an alleged violator may agree to undertake an environmentally beneficial project related to the violation in exchange for mitigation of the penalty to be paid. A Supplemental Environmental Project (SEP) furthers EPA's goal of protecting and enhancing public health and the environment. It does not include the activities a violator must take to return to compliance with the law.

[8] An enforcement action against American Electric Power resulted in a settlement between the federal government and the Ohio-based utility in October 2007. EPA officials said this particular estimate was conservative because it excluded, for example, the increased operations and maintenance costs of the plants and consideration of additional plants covered in the consent decree that would require retrofitting with pollution controls, conversion to different power sources, or retirement, which could cost more than $1 billion.

[9] In commenting on our draft report, EPA said that their decision was further informed by pollution control planning documents obtained through discovery, representations made in the litigation, and transcribed statements made to the federal court judge supervising confidential settlement negotiations.

[10] Discovery is the process where civil litigants seek and obtain information both from other parties to the litigation and others through, for example, interrogatories and document requests.

[11] 28 U.S.C. §§ 515-519.

[12] EPA, Office of the Inspector General, *New Source Review Rule Change Harms EPA's Ability to Enforce Against Coal-fired Electric Utilitie*s, 2004-P-0034 (Washington, D.C.: Sept. 30, 2004).

[13] After EPA issued the final New Source Review Equipment Replacement rule, 14 states, plus other governmental entities, and several public health/environmental organizations filed suits in the Court of Appeals for the District of Columbia Circuit challenging the rule. Some of these groups asked the Court to prevent the rule from taking effect or "stay the rule" until the challenges they raised in their lawsuits were resolved by the Court. On December 24, 2003, the Court stayed the effective date of the October 2003 Equipment Replacement New Source Review rule until the case could be fully adjudicated. As a result, the rule would not become effective on December 26, 2003. In March 2006, the U.S. Court of Appeals for the District of Columbia vacated the revised rule.

In: Enforcing Federal Pollution Control Laws
Editor: Norbert Forgács

ISBN: 978-1-60876-082-4
© 2010 Nova Science Publishers, Inc.

Chapter 3

ENVIRONMENTAL PROTECTION: EPA-STATE ENFORCEMENT PARTNERSHIP HAS IMPROVED, BUT EPA'S OVERSIGHT NEEDS FURTHER ENHANCEMENT[*]

United States Government Accountability Office

WHY GAO DID THIS STUDY

The Environmental Protection Agency (EPA) enforces the nation's environmental laws through its Office of Enforcement and Compliance Assurance (OECA). OECA sets overall enforcement policies and through its 10 regions oversees state agencies authorized to implement environmental programs consistent with federal requirements. GAO was asked to (1) identify trends in federal resources to regions and states for enforcement between 1997 and 2006, and determine regions' and states' views on the adequacy of these resources; (2) determine EPA's progress in improving priority setting and enforcement planning with states; and (3) examine EPA's efforts to improve oversight of states' enforcement programs and identify additional actions EPA could take to ensure more consistent state performance and oversight. GAO examined information from all 10 regions and 10 authorized states, among other things.

WHAT GAO RECOMMENDS

GAO recommends that EPA (1) develop an action plan to address problems identified in state programs, (2) evaluate the capacity of state programs to enforce authorized programs, (3) publish findings of state enforcement program reviews, and (4) assess the performance of

---

[*] This is an edited, reformatted and augmented version of a U. S. Government Accountability Office publication, Report GAO-07-883, dated July 2007.

its 10 regions. EPA generally agreed with GAO's recommendations, but stated it will decide whether to publish future state reviews when it evaluates the review process in fiscal year 2008.

WHAT GAO FOUND

Overall funding to regions and authorized states increased from 1997 through 2006, but these increases did not keep pace with inflation and the growth in enforcement responsibilities. Over the 10-year period, EPA's enforcement funding to the regions decreased 8 percent in inflation-adjusted terms. Regional officials said they reduced the number of enforcement staff by about 5 percent. EPA's grants to states to implement federal environmental programs also declined by 9 percent in inflation-adjusted terms while enforcement and other environmental program responsibilities increased. According to state officials, reductions in grant funds have limited their ability to meet EPA's requests to implement new requirements. For example, according to New York State officials responsible for the hazardous waste program, a reduction in EPA grants between 1997 and 2006 has meant a 38 percent reduction in the full-time state staff supported by federal funding for this program. However, EPA information on the workload and staffing needs of its regions and the states is incomplete, and, thus, it is not possible with existing data to determine their overall capacity to meet their enforcement responsibilities.

EPA has made substantial progress in improving priority setting and enforcement planning with states through its system for setting national enforcement priorities and the EPA/state National Environmental Partnership System (NEPPS), which have fostered a more cooperative relationship. For example, on states' recommendation, OECA accepted as a priority ensuring that facilities handling hazardous substances, such as lead or mercury, have the financial resources to close their facilities, clean up contamination, and compensate communities and individuals affected by the contamination. EPA and states have also made some progress in using NEPPS for joint planning and resource allocation. State participation in the partnership grew from 6 pilot states in fiscal year 1996 to 41 states in fiscal year 2006.

EPA has improved its oversight of state enforcement programs by implementing the State Review Framework (SRF) as a means to perform a consistent approach for overseeing the programs. Moreover, EPA can make additional progress by addressing weaknesses that the SRF reviews identified and by implementing other improvements to ensure oversight that is more consistent. For example, the SRF reviews show that EPA has limited ability to determine whether the states are performing timely, appropriate enforcement and whether penalties are applied to environmental violators in a fair and consistent manner within and among the states. In addition, GAO noted that EPA could make further use of the SRF to (1) determine the root causes of poorly performing programs; (2) inform the public about how well the states are implementing their enforcement responsibilities; and (3) extend the use of the SRF methodology to assess the performance of EPA's regions, which have been inconsistent in their enforcement and oversight efforts.

ABBREVIATIONS

EPA	Environmental Protection Agency
ECOS	Environmental Council of the States
FTE	Full-Time equivalent
NEPPS	National Environmental Partnership System
NYSDEC	New York State Department of Environmental Conservation
OECA	Office of Enforcement and Compliance Assurance
PPA	Performance Partnership Agreement
PPG	Performance Partnership Grant
RCRA	Resource Conservation and Recovery Act
SRF	State Review Framework
UST	Underground Storage Tank

July 31, 2007

The Honorable Norman D. Dicks
Chairman
The Honorable Todd Tiahrt
Ranking Member
Subcommittee on Interior, Environment and Related Agencies
Committee on Appropriations
House of Representatives

The Honorable James M. Inhofe
Ranking Member
Committee on Environment and Public Works
United States Senate

The Environmental Protection Agency (EPA), in partnership with state agencies, oversees compliance with 44 separate environmental programs. These programs regulate facilities—such as sewage treatment plants, petroleum refineries, and power plants—whose operations could pollute the air, water, and land, and thereby endanger public health and the environment. EPA and its regulatory partners are responsible for ensuring that these regulated facilities comply with program requirements and taking enforcement action in instances of noncompliance. These enforcement efforts are important for ensuring a level playing field because, among other things, facilities that do not comply with program requirements might have a competitive economic advantage over facilities that take environmental requirements seriously and thereby incur additional operational costs.

Many federal environmental statutes, such as the Clean Air Act, the Clean Water Act, and the Solid Waste Disposal Act, direct EPA to approve or authorize qualified states to implement and enforce environmental programs consistent with federal requirements. EPA expects its 10 regional offices to take a systematic, consistent approach in overseeing the state enforcement programs and, in doing so, to follow EPA's regulations, policies, and guidance.

EPA outlines, by policy and guidance, its oversight expectations for regional offices with regard to ensuring the state approaches include the elements of an acceptable state enforcement program, such as the type and timing of the actions that should be taken for various violations, and track how well the states comply.

This model of state enforcement of environmental laws, accompanied by EPA's regional oversight, allows the level of government closest to environmental conditions to assume primary responsibility for implementing programs. But it requires that states acquire and maintain adequate capacity to enforce state environmental programs that are consistent with federal requirements and act in a timely and appropriate manner to ensure violators come into compliance. EPA establishes by regulation the requirements for state enforcement authority, such as the authority to seek civil and criminal penalties and injunctive relief.[1] EPA grants authorization to the states on a program-by-program basis. EPA policy and guidance outline, with greater specificity, the elements of an acceptable state enforcement program—such as the necessary legislative authorities and the type and timing of the enforcement actions for various violations—and track how well states comply.

Most states have responsibility for multiple EPA programs. EPA-authorized states monitor the compliance of regulated facilities by conducting inspections, performing evaluations, and reviewing records to verify facilities' compliance with programs regulating the discharge of pollutants into surface water or the air and the storage and disposal of hazardous waste. States are expected to pursue enforcement actions against those facilities found in noncompliance and to report their actions to EPA.

EPA administers its environmental enforcement responsibilities through its headquarters Office of Enforcement and Compliance Assurance (OECA). OECA monitors the compliance of regulated facilities, identifies national enforcement concerns and sets priorities, and provides overall direction on enforcement policies. While OECA headquarters occasionally takes direct enforcement action, much of EPA's enforcement responsibilities are carried out by its 10 regional offices. These offices are responsible for carrying out core program activities under each of the major federal environmental statutes, as well as significant involvement in implementing EPA's national enforcement priorities and taking direct enforcement action. The regions are also responsible for overseeing authorized states' enforcement programs, implementing programs in Indian country[2] and states that are not authorized for particular programs. OECA also expects regions and authorized states to establish enforcement priorities and expectations and reach agreement on their respective roles and responsibilities. Authorized states may also receive EPA grants to assist in implementing and enforcing authorized programs. In fiscal year 2006, grants to states and tribes totaled $3.2 billion, or about 42 percent of EPA's total budget.

Over the years, states have increased their inspection and enforcement activities. As a result, EPA regional offices are now more actively involved in conducting oversight and providing states with guidance, training, and technical assistance to assure consistent performance of state enforcement programs. If EPA finds a state is not adequately administering or enforcing authorized programs, individual environmental statutes may authorize EPA to take certain actions, including providing additional technical assistance, conditioning the receipt of grant funds on compliance with EPA guidance, or withdrawing state authorization. In addition, when EPA finds a specific state enforcement action to be inadequate, the agency may take federal enforcement action against the violator.

Despite the interdependence between EPA and the states in carrying out enforcement responsibilities, effective working relationships have historically been difficult to establish and maintain, as we, EPA's Office of Inspector General, the National Academy of Public Administration, and others have reported.[3] The following three key issues have affected EPA-state relationships:

- EPA's funding allocations to the states have not fully reflected the differences among the states' enforcement workload and their relative ability to enforce state environmental programs consistent with federal requirements. In this regard, EPA lacks information on the capacity of both the states and EPA's regions to effectively carry out their enforcement programs, because the agency has done little to assess the overall enforcement workload of the states and regions and the number and skills of people needed to implement enforcement tasks, duties, and responsibilities. Furthermore, the states' capacity continues to evolve as they assume a greater role in the day-to-day management of enforcement activities, workload changes occur as a result of new environmental legislation, new technologies are introduced, and state populations shift.
- Problems in EPA's enforcement planning and priority setting processes have resulted in misunderstandings between OECA, regional offices, and the states regarding their respective enforcement roles, responsibilities, and priorities. States have raised concerns that EPA sometimes "micromanages" state programs without explaining its reasons for doing so and often does not adequately consult the states before making decisions affecting them.
- OECA has not established a consistent national strategy for overseeing states' enforcement of EPA programs. Consequently, the regional offices have not been consistent in how they oversee the states. Some regional offices conducted more in-depth state reviews than others, and states in these regions have raised concerns that their regulated facilities are held to differing standards of compliance than facilities in states located in other regions.

EPA and leaders of state environmental programs have tried over the years to establish new mechanisms to address each of these long-standing problems in order to strengthen the EPA-state partnership. For example, EPA has linked its budgeting and allocation process to its strategic goals and objectives, and makes strategic decisions in developing its budget for its enforcement workforce to reflect shifting priorities. Nonetheless, according to OECA officials, shifts in funding and staff years made as a result of changing priorities are generally marginal. In this regard, in July 2005 we reported that an effective workforce strategy is needed, particularly during times of fiscal constraint, so that OECA can tailor workforce changes to reflect actual conditions in the regions and states and minimize potential adverse impacts on EPA's programs.[4]

To better clarify roles, responsibilities, and priorities between EPA and the states, the agency established the National Environmental Performance Partnership System (NEPPS) in 1995 to give states demonstrating strong environmental performance greater flexibility and autonomy in planning and operating their environmental programs. Under this system, a state and EPA may enter into a Performance Partnership Agreement (PPA) that identifies the state's environmental goals and priorities, and spells out how EPA and state officials are to address them. States may also ask to combine EPA grants into a Performance Partnership Grant (PPG), which is intended to allow the state greater flexibility in targeting limited resources to meet its priorities. In 2003, OECA revised its process for setting national enforcement priorities to better consider the views of states and regions and to more effectively target enforcement resources. Under the new process, OECA headquarters evaluates the overall environmental performance of individual industrial sectors, and solicits views about those sectors from states and regions, the representatives of the industrial sectors, and the public. OECA uses this information to identify priorities that are best addressed through focused federal attention. OECA issues guidance to the regions and states for implementing the national enforcement priorities, as well as guidance for deterring noncompliance among all regulated sectors.

Likewise, OECA implemented a new oversight program in 2004, known as the State Review Framework (SRF), to more uniformly and objectively measure the performance of states' enforcement programs. Under SRF, regions evaluate the extent to which state performance in managing three major programs complies with specific legal requirements, policy, and guidance, while OECA headquarters manages the overall review process.[5] In conducting this evaluation, the regions use 12 review elements, such as the degree to which states complete planned inspections, accurately identify significant violations, and take timely and appropriate enforcement action. As of January 2007, EPA had conducted SRF reviews in 33 states and expected to complete its assessment of the remaining states by the end of fiscal year 2007. EPA plans to evaluate the implementation of SRF in fiscal year 2008. Appendix II contains a detailed description of the development of the SRF.

In this context, you asked that we (1) identify trends in the federal resources provided to regions and states for enforcement between 1997 and 2006, and determine regional and states' views on the adequacy of these resources to implement their activities; (2) determine EPA's progress in improving priority setting and enforcement planning with its regions and authorized states; and (3) examine EPA's efforts to improve its oversight of state's enforcement and compliance programs and identify additional actions that could be taken to ensure more consistent state performance and more consistent oversight of state programs.

To address these issues, we reviewed EPA's strategic plans and national strategy, and its policy and guidance for planning and implementing its enforcement programs and for establishing performance partnerships with authorized state agencies.[6] We also examined the budgets for EPA and OECA for fiscal years 1997 through 2006. We discussed the development and implementation of national strategy, policy, guidance, and resource allocation with officials of OECA, EPA's Office of Congressional and Intergovernmental Relations, EPA's Office of Chief Financial Officer, and officials responsible for state program oversight in each of EPA's 10 regional offices. In each region, we examined regional strategic plans, partnership agreements with authorized states, and state oversight reviews. We used semi-structured interviews to elicit, organize, and evaluate narrative responses from officials at the 10 regional offices and 10 authorized state agencies. We selected one state from each region: five states that had performance partnership agreements with EPA (Iowa, Massachusetts, Minnesota, Oregon, and Utah) and five that did not (Arizona, Arkansas, Florida, New York, and West Virginia). At these state agencies, we reviewed strategic plans and strategy, and policy and guidance for planning and implementing enforcement programs. We were not able to assess the workload and capacity of states to meet their programs because, as mentioned previously, EPA does not have a system to collect information needed

Environmental Protection 57

for such an assessment, including consistent and complete information from the states or regions on their workload and the number, type, and skills of staff needed to carry out these responsibilities. Appendix I contains a detailed description of our scope and methodology. We performed our work from October 2005 through July 2007 in accordance with generally accepted government auditing standards, which included an assessment of data reliability and internal controls.

RESULTS IN BRIEF

Overall funding to regions and authorized states increased from fiscal years 1997 through 2006. However, these increases did not keep pace with inflation and the growth in enforcement responsibilities. Both EPA and state officials told us they are finding it difficult to respond to new requirements while carrying out their previous responsibilities. Over the 10-year period, EPA's enforcement funding to the regions increased from $288 million in fiscal year 1997 to $322 million in fiscal year 2006, but declined in real terms by 8 percent. In response, regional officials said, they reduced the number of enforcement staff by about 5 percent. EPA's grants to states to implement environmental programs consistent with federal requirements also increased over the 10-year period, from $2.9 billion in fiscal year 1997 to $3.2 billion in fiscal year 2006, but declined in real terms by 9 percent. In addition, grant funding dropped substantially between fiscal years 2004 and 2006, from $3.9 billion to $3.2 billion. These reductions in funding occurred during a period when statutory and regulatory changes increased enforcement and other environmental program responsibilities. As we reported in July 2005, EPA's implementation of amendments to the Clean Water Act (1) increased the number of regulated industrial and municipal facilities by an estimated 186,000 facilities and (2) added hundreds of thousands of construction projects to states' and regions' workloads for the storm water program. For example, Arkansas officials said that, for the storm water program in their state, the number of inspections and storm water permits issued increased by 512 percent from 2003 to 2005. In addition, other state officials told us that reductions in grant funds have limited their ability to meet EPA's requests for their states to carry out new requirements. As a result, as these officials focus on priority enforcement work, they have accumulated a backlog of work in other areas, such as renewing permits for regulated facilities. However, EPA does not collect sufficient information on enforcement workload and staffing to permit an independent assessment of the capability of either regions or states to meet their enforcement responsibilities.

EPA has made substantial progress in improving priority setting and enforcement planning with states through its system for setting national enforcement priorities and NEPPS, which have fostered a more cooperative relationship. With respect to setting national priorities, for example, stakeholders recommended, and OECA accepted, a focus on ensuring that facilities handling hazardous substances, such as lead and mercury, have the financial resources necessary to close their facilities, clean up any contamination, and compensate communities and individuals affected by the contamination. Because financial assurance is now a national priority, EPA will provide training to state inspectors, who often do not have a financial background, on how to assess the adequacy of documented financial resources. EPA and states have also made progress in using NEPPS for establishing enforcement responsibilities and allocating resources. State participation in the partnership grew from 6 pilot states in fiscal year 1996 to 41 states in fiscal year 2006, although the extent of participation in the partnership varies. Of the 41 states, 31 had both Performance Partnership Agreements and Performance Partnership Grants; 2 had agreements only; and 8 had grants only. Participation has been uneven, according to state officials, because the benefits expected from this partnership—flexibility in managing their programs and directing grant funds—have only been partially realized. For example, Minnesota and Massachusetts officials said they have been able to achieve some flexibility from EPA to target state priorities and allocate grant resources to those priorities. Other states, such as Florida and New York, reported that they did not see significant benefits from participation in NEPPS. For example, Florida officials told us that state appropriation procedures restrict their ability to shift resources among programs.

EPA has improved its oversight of state enforcement programs by implementing the SRF to provide a consistent approach for overseeing the programs. Moreover, EPA can make additional progress by addressing weaknesses that the SRF reviews identified and by implementing other improvements to ensure oversight that is more consistent. With its implementation of the SRF, EPA has, for the first time, a consistent approach for overseeing states' compliance and enforcement programs. The SRF reviews have also identified several significant weaknesses in how states enforce program requirements. For example, the reviews frequently found that states are not properly documenting inspection findings or penalties, as directed by EPA's enforcement policy and guidance. While recognizing that these findings are useful, EPA has not developed a plan for how it will uniformly address them in a timely manner. Nor has the agency identified the root causes of the weaknesses, although some EPA and state officials attribute the weaknesses to causes such as increased workloads concomitant with budgetary reductions. Until EPA addresses enforcement weaknesses and their causes, it faces limitations in determining whether states perform timely and appropriate enforcement and apply penalties to environmental violators in a fair and consistent manner within and among the states. The SRF is still in its early stages of implementation and offers additional uses that EPA has not yet considered, such as using the SRF's structured approach to (1) provide consistent information to the public on how well the states are implementing their enforcement responsibilities and (2) serve as a basis for assessing the performance of EPA's regions, which have been inconsistent in their enforcement and oversight efforts in the past.

In order to enhance EPA's oversight of regional and state enforcement activities consistent with federal requirements, we are recommending that the Administrator of EPA use the results of the SRF to (1) identify lessons learned and develop an action plan to address significant problems; (2) address capacity issues, such as state staffing levels and workload requirements, of state programs that perform poorly; (3) publish the results of the SRF so that the general public and others will know how well state regulators are enforcing authorized programs; and (4) for regional enforcement programs, conduct a performance assessment similar to the SRF. In commenting on the draft report, EPA generally agreed with GAO's recommendations and stated that the agency is taking action to address the issues we raised. With respect to our recommendation to publish the results of the SRF findings, EPA said that it had agreed with the Environmental Council of the States that the first round of state enforcement reviews would not be published. However, EPA said it would consider whether to publish future reviews when it evaluates the implementation of the SRF in fiscal year 2008.

BACKGROUND

Since its creation in 1970, EPA has generally been responsible for ensuring the enforcement of the nation's environmental laws. This responsibility has traditionally involved monitoring compliance by those in the regulated community (such as factories or businesses that release pollutants into the environment or use hazardous chemicals), ensuring that violations are properly identified and reported, and ensuring that timely and appropriate enforcement actions are taken against violators when necessary.

Because states historically had primary responsibility for addressing environmental pollution, including taking the lead role in enforcement under federal environmental legislation of the 1950s and 1960s, many states were dissatisfied with the new enforcement powers Congress granted to EPA. Referring to his first term as EPA Administrator during the 1970s, William Ruckelshaus described relations between EPA and state governments as "terrible," largely because EPA itself represented a repudiation of what the state regulators had been doing. The states felt, he believed, that in the face of very little public or political support, they had made considerable progress and were getting no credit for it. The very existence of EPA symbolized to state environmental agencies the lack of appreciation the public had for their efforts and accomplishments.

Furthermore, because EPA was now responsible for ensuring the implementation and enforcement of environmental laws, states needed to demonstrate to EPA that they had acquired and were maintaining adequate authority to enforce program requirements consistent with federal law before EPA would "authorize" and continue to allow states to assume the day-to-day administration of new environmental programs. This federal oversight contributed to a very difficult period between EPA and the states. The states thought EPA dictated too much and was too intrusive.

Difficulties in the EPA-state relationship manifested themselves from the 1970s through the 1990s and, to some degree, have continued to the present day. In 1980, we described the failings of the EPA-state partnership.[7] In a 1988 comprehensive management review of EPA, we reported that while some progress had been made in improving the EPA/state relationship, the goal of a truly effective EPA-state partnership remained elusive.[8] Many state officials expressed concerns about having limited flexibility, too much EPA control, and excessively detailed EPA oversight.

In 2003, OECA officials said that environmental commissioners in several states and members of the Environmental Council of the States (ECOS)[9] called for OECA to develop and implement a consistent and objective mechanism for measuring state performance. Specifically, ECOS wanted EPA's oversight of state programs to be consistent between EPA regions and states in the same region. ECOS also wanted EPA oversight to be predictable, repeatable, and unbiased. In December 2003, OECA began jointly developing the SRF with input from EPA regions, associations representing state pollution control agencies,[10] ECOS and state officials to evaluate the extent to which state performance in three major programs complies with specific legal requirements, policy and guidance.[11] The SRF measures state performance using 12 required elements and an optional 13th element. The elements include five categories: (1) review of state inspection implementation; (2) review of state enforcement activity; (3) review of state enforcement commitments; (4) review of state data integrity; and (5) for the optional 13th element, a review of additional programs of the state's choice to

60   United States Government Accountability Office

insure consideration of state activities that support the overall evaluation. Regions prepare a draft report of the findings and conclusions of each review, jointly discuss with the state how major recommendations will be addressed, and provide the draft reports to OECA headquarters. OECA reviews the draft reports and provides comments on the regions' analyses and recommendations. A more detailed description of the SRF is included in appendix II.

The relationship between EPA and state environmental agencies varies substantially from state to state and program to program. Staff in EPA's regional offices operate programs in those states that have elected not to seek authorization and in those states that EPA has concluded are not prepared to manage the programs effectively. In some cases, where EPA has denied a state with program authorization, state personnel may do most of the work as if they had the authority to grant permits, with EPA employees handling only the final step of formally issuing the documents. In other states or programs, EPA may approve a "partial delegation" of authority, under which a state operates part of a program that can be delegated and EPA operates the remainder.

Just as EPA can authorize a state to conduct day-to-day program management, environmental statutes may allow EPA to withdraw authorization if a state fails to meet certain conditions, including maintaining the capacity to effectively manage the program and adopting and properly exercising the legal authorities to enforce program compliance consistent with federal laws and regulations. In practical terms, however, EPA's ability to directly manage state programs is limited because of its staffing levels and other resources.

INCREASES IN EPA FUNDING HAVE NOT KEPT PACE WITH INFLATION AND ENFORCEMENT RESPONSIBILITIES

Overall funding to regions and authorized states increased from fiscal years 1997 through 2006. However, these increases did not keep pace with inflation and the growth in enforcement responsibilities. Both EPA and state officials told us they are finding it difficult to respond to new requirements while carrying out their previous responsibilities. However, EPA does not collect sufficient data on enforcement workload and staffing, which it needs to assess the capacity of the regions and states to effectively implement their responsibilities for enforcing environmental laws consistent with federal requirements.

Resources for EPA Regions and the States Have Declined in Real Terms

According to our analysis of EPA's budget and workforce for fiscal years 1997 through 2006, EPA's total budget increased from $7.3 billion to $7.7 billion—a decline of 13 percent in real terms.[12] At the same time, total funding for EPA enforcement increased from $455 million to $522 million—a decline of 5 percent in real terms. For the regions, which command the bulk of enforcement resources, funding increased from $288 million to $322 million—a decline of 8 percent in real terms, while headquarters enforcement funding increased from $167 million to $200 million—a decline of 1 percent in real terms. Figure 1 shows the changes in EPA enforcement funding in real terms, in total, and by headquarters and regional offices.

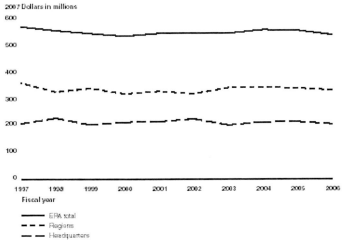

Source: GAO analysis of EPA data.
Note: In nominal terms, funding enforcement in EPA regions increased from $288 million in fiscal year 1997 to $322 million in fiscal year 2006.

Figure 1. EPA's Enforcement Funding in Real Terms, in Total, and by Headquarters and Regional Offices, Fiscal Years 1997-2006.

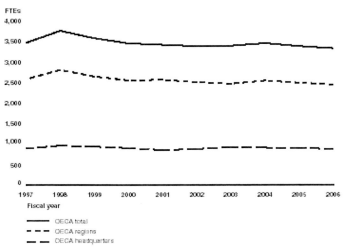

Source: EPA Office of Budget.

Figure 2. OECA Headquarters and Regional Total FTEs, Fiscal Years 1997-2006.

According to officials in OECA and EPA's Office of the Chief Financial Officer, OECA headquarters absorbed decreases in OECA's total enforcement funding in recent years to prevent further reductions to the regions. According to our analysis, enforcement funding for OECA headquarters increased from $197 million in fiscal year 2002 to $200 million in fiscal year 2006—a 9 percent decline in real terms. During the same time, regional enforcement funding increased from $279 million to $322 million—a 4 percent increase in real terms.

EPA reduced the size of the regional enforcement workforce by about 5 percent over the 10 years, from 2,568 full-time equivalent (FTE) staff in fiscal year 1997 to 2,434 FTEs in fiscal year 2006. In comparison, the OECA headquarters workforce declined 1 percent, and the EPA total workforce increased 1 percent during the same period. Figure 2 shows the changes in headquarters and regional FTEs from fiscal years 1997 through 2006.

As figure 3 shows, the change in FTEs was not uniform across the 10 regions over the period. For example:

- Two regions—Region 9 (San Francisco) and Region 10 (Seattle)— experienced increases in their workforce: Region 9 increased 5 percent, from 229 to 242 FTEs, and Region 10 increased 6 percent, from 161 to 170 FTEs.
- Two regions—Region 1 (Boston) and Region 2 (New York) experienced the largest declines: Region 1 experienced a 15 percent decline, from 195 to 166 FTEs, and Region 2 had a 13 percent decline, from 291 to 254 FTEs.

As EPA's real total funding declined, EPA's real total grant funding to states and tribes declined, as shown in figure 4. States and tribes use these grant funds, combined with their own resources, to implement and enforce environmental programs consistent with federal requirements. EPA's grants to authorized states and tribes increased from $2.9 billion to $3.2 billion from fiscal years 1997 through 2006—a decline of 9 percent in real terms. However, grant funding to states and tribes dropped substantially between fiscal years 2004 and 2006, from $3.9 billion to $3.2 billion—a 22 percent decline in real terms.

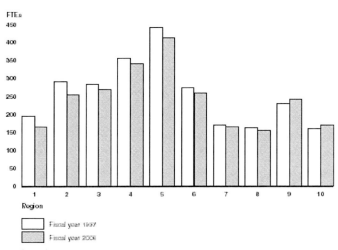

Source: EPA Office of Budget.

Figure 3. OECA Regional Offices FTEs, Fiscal Years 1997-2006.

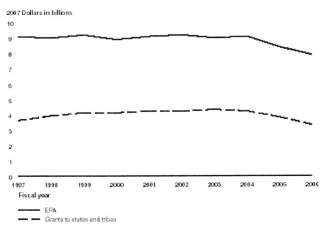

Source: GAO analysis of EPA data.

Note: In nominal terms, funding for grants to states and tribes increased from $2.9 billion in fiscal year 1997 to $3.2 billion in fiscal year 2006.

Figure 4. EPA's Real Total Funding and Real Total Grant Funding to States and Tribes, Fiscal Years 1997 through 2006.

For the programs we examined, EPA provides states with federal resources through grant programs. In addition, under the Clean Air Act, states must collect fees from permitted facilities to help fund the program. The following are some of the funding sources for the programs we examined:

- Clean Water Act Section 106 Grants. The Office of Water provides annual Water Pollution Control grant funds to the states under the Clean Water Act for, among other things, enforcement of point source pollution requirements.

- Clean Air Act Section 105 Grants. States finance enforcement of air quality laws with fees paid by permitted facilities (e.g., electric utilities and chemical manufacturers), in accordance with title V of the Clean Air Act. States assess fees based on tons of pollution emitted. The fee must be at least $25 per ton (as adjusted for inflation) of regulated pollutants (excluding carbon monoxide). EPA provides grants to states to help fund programs that prevent and control air pollution or to implement national ambient air quality standards, including programs that may address smaller sources that do not have to obtain permits (e.g., dry cleaners and gasoline stations).

- Resource Conservation and Recovery Act (RCRA) Section 3011 Grants. The Office of Solid Waste and Emergency Response provides annual grants to states under section 3011 of the Solid Waste Disposal Act to support, among other things, state enforcement of the RCRA Subtitle C Hazardous Waste Management programs.13 EPA's financial assistance covers up to 75 percent of a state's total costs for managing hazardous waste.

64   United States Government Accountability Office

- Underground Storage Tank (UST) Grants. The Office of Underground Storage Tanks provides grants to states under subtitle I of RCRA to support state UST programs, including inspections and enforcement activities.

EPA Regions and States Find It Difficult to Respond to New Requirements under Current Resource Constraints

Given the real reductions in funding and personnel, regional and state enforcement officials noted states are finding it difficult to respond to new enforcement requirements in the Clean Water Act, Clean Air Act, and RCRA, which have greatly increased the number of regulated pollutants and sources.[14] However, as we reported in June 2006, EPA's data collection system for state and regional enforcement activities does not provide consistent and complete data on the workload of the states or the regions, and on their capacity to meet their workload, including the enforcement of environmental programs consistent with federal requirements.[15] Thus, we were not able to independently assess their capabilities to meet their responsibilities under environmental programs with their existing resources.

Clean Water Act

In 1987 amendments to the Clean Water Act,[16] Congress expanded the scope of the act by regulating storm water runoff from rain or snow as a discharge from point sources, such as industrial facilities and municipal separate storm sewer systems. We reported in 2005 that that the new storm water regulations increased the number of industrial and municipal facilities subject to regulation by an estimated 186,000 facilities. In that same year, the EPA Office of Inspector General reported that the new requirements added hundreds of thousands of construction projects to states' and regions' workloads related to storm water pollution sources. In 2005, EPA established storm water as a national enforcement priority.

According to state water quality enforcement officials, it is difficult to meet these new requirements while still meeting older requirements to periodically inspect large municipal and industrial point sources. For example, Minnesota officials told us that there are a huge number of facilities and locations that require inspection and permits in that state. The Minnesota storm water officials provided us a performance report that indicated noncompliance with storm water requirements was so pervasive that the state's water enforcement program was not meeting its timeliness targets for closing enforcement cases where it has the authority to issue a penalty. The report said that the situation threatened to undermine earlier success in implementing the state's storm water permit program. According to the Arkansas director of the state's water quality program, storm water enforcement work in the fast-growing northwestern corner of the state has overwhelmed the program. According to state data, the number of storm water permits issued grew by 512 percent—from 427 to 2,616—between 2003 and 2005. At the same time, the number of permits Arkansas officials issued to traditional point sources increased by 19 percent, from 190 to 227 facilities. However, the state is beginning to fall behind in issuing the traditional permits because it has had to redeploy inspectors to work on storm water enforcement.

Regional officials told us they have provided assistance to some states to meet enforcement requirements for storm water, such as helping states complete a specified number or percentage of state inspections. For example, Region 8 (Denver) enforcement officials said they share the inspection workload with Montana officials because that state's permit program is chronically underfunded, and this partnership enables state officials in Montana to focus on reducing their backlog of regulated facilities that require permit renewals.

Clean Air Act

In 1990 amendments to the Clean Air Act,[17] Congress created significant new enforcement requirements for both EPA regions and states.[18] Title III of the 1990 amendments required EPA to control emissions of 189 air toxics by, among other things, developing technology-based emissions limits for major stationary sources.[19] By 2006, there were an estimated 84,000 major stationary sources within 158 major source categories such as incinerators and chemical plants to which these standards applied. Title III of the amendments also directed EPA to develop a list of categories of small stationary sources, such as dry cleaners and gas stations—so-called "area sources"—sufficient to represent 90 percent of emissions from small stationary sources of listed hazardous air pollutants. EPA and authorized states were then to implement strategies to control toxic emissions from these sources. EPA developed a list of 70 categories, but, as we reported in 2006, it had issued standards for only 16.[20]

Under title V of the 1990 amendments, Congress directed EPA and authorized states to implement a comprehensive permit program for sources emitting regulated air pollutants.[21] The new requirement significantly expanded an earlier permit program that applied only to major new construction or modifications of existing major sources of pollution. As of April 2001, EPA had authorized all 50 states and 63 local governmental units to act as clean air regulatory agencies and issue Title V operating permits. The 1990 amendments also required states to collect fees from regulated facilities sufficient to cover program costs. We reported in 2001 that 19,880 major sources had already received, or could be expected to obtain or comply with the conditions of, Title V permits;[22] furthermore, federal and state regulators performed about 17,800 routine inspections each year in fiscal years 1998 and 1999.

State air enforcement officials said that reductions in federal air quality grants and EPA restrictions on how states can use the permit fees limit their ability to meet new Clean Air Act requirements. For example, enforcement officials in Arizona and West Virginia told us that reductions in section 105 grants have made it difficult to meet EPA's push for states to identify and regulate air toxic emissions from small stationary sources, such as dry cleaners. According to the director of West Virginia's air quality program, while his office has identified these sources and provided this information to Region 3 (Philadelphia), it does not have the resources to regulate these sources. Title V rules restrict states from using permit fees for enforcement of air toxic emission standards for small stationary sources.

Several state officials said they also face difficulties keeping up with Title V requirements to issue operating permits and inspect facilities. For example, according to West Virginia's Director of Air Quality, the state program is spending more than it generates in revenues from Title V permit fees. Furthermore, the state is drawing down a reserve fund that it established in 1993, when it was operating a preliminary Title V program before it was authorized to administer this program. West Virginia's Title V fee is around $21 per ton, which is below the national average fee of $28 per ton paid by major sources in 2001.[23] The Director of Arizona's Air Quality Division said that as a result of new standards and new types of performance testing, the Title V permitting workload has increased faster than staff levels.

Resource Conservation and Recovery Act

The 1984 amendments to RCRA[24] required EPA and authorized states to regulate small-quantity hazardous waste generators—those producing between 100 and 1,000 kilograms of waste per month. Before the 1984 amendments, EPA only required generators who produce more than 1,000 kilograms of hazardous waste to comply with regulations concerning, among other things, record keeping and reporting. In addition, the 1984 amendments required EPA and authorized states to ensure that owners and operators of treatment, storage and disposal facilities comply with new prohibitions for disposing of untreated hazardous wastes on land.

This expansion of responsibility, combined with a reduction in resources, has made it difficult to carry out responsibilities, according to state hazardous waste enforcement officials we spoke with. For example, data provided to us by New York State officials showed that the state's grant for hazardous waste management dropped in real terms by 19 percent between fiscal years 1997 and 2006. As a result of this decline in funding, New York State officials said that the grant supports fewer full-time staff. The officials provided us data that showed New York State's grant for hazardous waste management supported 38 percent fewer full-time staff in the state's RCRA program in fiscal year 2006 than in fiscal year 1997. Furthermore, while it continues to focus on EPA's national priorities for enforcement, the state has accumulated a backlog of permits that must be renewed. Renewing these hazardous waste permits is critical to protecting the environment and public health because EPA and authorized states can enforce new hazardous waste standards only when they are specified in a permit. According to these officials, EPA has issued new standards during the life of the old permit and these new standards are not enforceable until the permit is renewed.

Several states identified an emerging challenge: verifying that owners and operators of hazardous waste facilities have the financial resources to clean up their sites, as RCRA requires, rather than leaving site cleanup to the taxpayer. As we reported in 2005, EPA's Office of Inspector General concluded that required cleanups at some sites could exceed $50 million. EPA made enforcement of the financial assurance requirement a national enforcement priority in 2005. However, state officials told us, and EPA's regional officials agreed, that they do not have the personnel with the necessary skills to evaluate the financial assurances provided. For example, New York State officials told us that they have only one person trained to review the adequacy of a facility's financial assurance.

The Energy Policy Act of 2005[25] also expanded states' enforcement workload by requiring, among other things, (1) EPA and any state receiving federal funding to inspect by August 8, 2007, all regulated tanks that were not inspected since December 22, 1998, and (2) EPA or the state must generally inspect regulated tanks once every 3 years and complete the first 3-year inspection cycle within 3 years of completing inspections of underground storage tanks that had not been inspected by December 22, 1998.[26] We reported in February 2007 that according to EPA data there were 645,990 active federally regulated underground storage tanks registered with state underground storage tank programs.[27]

To carry out these additional inspections, the act, as amended, allows states to use Leaking Underground Storage Tank Trust Fund appropriations, and authorizes substantial appropriations from the trust fund during fiscal years 2006 through 2011. These new appropriations were to be in addition to State and Tribal Grant funds EPA provides to states for underground storage tank programs, which historically have been about $187,000 annually. These additional funds were not appropriated in fiscal years 2006 and 2007. As a result, some state officials told us, they do not have sufficient resources to meet the new inspection requirements. For example, Minnesota officials said they have sufficient resources for inspecting underground storage tanks only once every 9 years. Regional officials acknowledged that many states need additional inspectors for meeting these new underground storage tank inspection requirements and said they provide support to states facing fiscal constraints. For example, the Region 3 Chief of RCRA enforcement and compliance said the region helped state officials in West Virginia by inspecting small and midrange gas stations in the state. Region 9 (San Francisco) has also been working with UST officials in Arizona to meet the new inspection requirements.

EPA AND STATES HAVE TAKEN STEPS TO IMPROVE PLANNING AND PRIORITY SETTING FOR ENFORCEMENT, BUT RESULTS ARE UNEVEN

EPA has made substantial progress in improving priority setting and enforcement planning with states through its system for setting national enforcement priorities and NEPPS, which have fostered a more cooperative relationship. For example, on states' recommendation, OECA accepted as a priority ensuring that facilities handling hazardous substances, such as lead or mercury, have the financial resources to close their facilities, clean up contamination, and compensate communities and individuals affected by the contamination. EPA and states have also made some progress in using NEPPS for joint planning and resource allocation. State participation in the partnership grew from 6 pilot states in fiscal year 1996 to 41 states in fiscal year 2006.

EPA Implemented a More Collaborative Process for Setting National Priorities for Enforcement

Since the late 1980s, EPA's top agency executives have set priorities for improving agency management. To develop more collaborative relationships among EPA's headquarters and regions, and the states, OECA created the Planning Council in 2003 to direct OECA's strategic and planning processes, including selecting national enforcement priorities. The council includes both EPA headquarters and regional officials. In the summer of 2003, OECA asked officials representing state environmental agencies, tribal governments, and air, water, and solid waste pollution control associations to recommend priorities for consideration as national priorities for fiscal years 2005 through 2007. Considering these recommendations, as well as those from EPA's media program managers, OECA developed a list of potential national priorities that it published in the *Federal Register* in December 2003. OECA asked for public comments on candidate priorities and suggestions for new national enforcement priorities. In January 2004, OECA hosted a meeting with headquarters and regional officials, and representatives from 10 state environmental agencies, four tribal governments, and three pollution control associations to discuss and rank their choices of potential priorities. OECA then selected the final national enforcement priorities for fiscal years 2005 through 2007 based on the following criteria:

- Significant environmental benefit. In what specific areas can the federal enforcement and compliance assurance programs make a significant positive impact on human health and/or the environment? What are the known or estimated public health or environmental risks?

- Noncompliance. Are there particular economic or industrial sectors, geographic areas, or facility operations where regulated entities have demonstrated serious patterns of noncompliance?

- EPA responsibility. What identified national problem areas or programs are better addressed through EPA's federal capability in enforcement or compliance assistance?

Table 1 shows the potential national priorities, and those that were selected.

Table 1. EPA's Candidate and Final National Enforcement Priorities for Fiscal Years 2005-2007, in Rank Order

Potential priorities	Priorities selected
Clean Air Act	
Air toxics	•
New source review/prevention of significant deterioration	•
Clean Water Act	
Effects of wet weather on concentrated animal feeding operations, combined sewer overflows, sanitary sewer overflows, and storm water runoff	•
Petroleum refining	•[a]
Resource Conservation and Recovery Act	
Financial assurance/financial responsibility	•
Mineral processing	•
Tribal environmental issues	•
Federal facilities	
Resource Conservation and Recovery Act	
Leaking underground storage tanks	
Ports of entry	
Asbestos in schools	
Auto salvage yards	
Environmental justice	
Miscellaneous plastics	
Safe Drinking Water Act	
Microbials	
Fuels management	
Significant noncompliance oversight	

Source: EPA.

[a] EPA's national petroleum refining priority addresses air emissions from the nation's petroleum refineries. EPA reached its goal of addressing 80 percent of the domestic refining capacity through settlements or filed civil actions. As a result, OECA removed it from the national priority list at the end of fiscal year 2006.

According to state and regional officials, OECA's approach for selecting national priorities has fostered a more collaborative working relationship, and they pointed to the financial assurance national priority under RCRA as a case in point. This priority focuses on ensuring that owners and operators of facilities handling hazardous substances, such as lead or mercury, provide assurance that they have the financial resources necessary to close their facilities, clean up any contamination, and compensate communities and individuals affected by any contamination they cause. Stakeholders—EPA regions, states, pollution control associations,[28] and the public—told OECA that financial assurance should be a national priority. EPA and state supporting investigations confirmed that (1) there are significant noncompliance issues relating to the financial assurance requirements; (2) the issues are of national importance and deal with several environmental laws and regulatory programs; and (3) areas of noncompliance were not isolated to a specific sector, industry, or geographic location. In selecting financial assurance as a national priority, OECA stated that an effective national enforcement and compliance strategy would help address many of the problems created by the regulated facilities' failure to fulfill their financial responsibility obligations. In this regard, EPA is providing training to state inspectors— who often do not have expertise in financial management—on how to assess the adequacy of financial documentation provided by regulated facilities.

While state and regional officials told us the priority setting process is more collaborative, some state officials raised a number of issues about how EPA considered their comments about national priorities. For example, Massachusetts and Oregon officials told us that they did not understand why OECA ranked some of their priorities lower than the national priorities that were selected and would like to have feedback on how their views were considered when selecting national enforcement priorities. In addition, while Minnesota, Utah, and Arizona officials agree that the planning and priority setting has improved, they would prefer earlier involvement in the decision-making process, that is, in developing the candidate list. On the other hand, New York and Arkansas said that they preferred to provide input on the national enforcement priorities through the pollution control associations and/or ECOS. They said, by doing so, OECA will receive states' collective views on the candidate priorities in a more representative and compelling way.

After OECA selects the national enforcement priorities, it uses the National Program Managers guidance to inform its stakeholders of the national priorities and required elements of all environmental laws. The managers establish overall national goals for their respective programs based on a variety of factors, including underlying statutory mandates, congressional directives, administration/administrator priorities, and their own view of programs and policies that the programs should focus on. EPA regions and states use this guidance to negotiate agreements based on (1) which environmental problems will receive priority attention within state programs, (2) what EPA's and the states' respective enforcement roles and responsibilities will be, and (3) how the states' progress in achieving program objectives will be assessed. The results of these negotiations are documented primarily in PPAs and/or PPGs. According to EPA, NEPPS allows regions and states to negotiate agreements that vary in content and emphasis to reflect regional and state conditions and priorities.

Extent of States' Participation in the Planning and Priority Setting Process Varies

State participation in NEPPS grew from 6 pilot states in fiscal year 1996 to 41 states in fiscal year 2006. Of these 41 states participating in NEPPS in fiscal year 2006, 31 had both PPAs and PPGs; 2 had agreements only; and 8 had grants only. Twelve states did not participate at all in NEPPS.

Regional officials and states participating in NEPPS said it contributed to improvements in their planning and priority setting process by helping identify enforcement priorities, roles, and responsibilities. The states we spoke with that had a PPA—Iowa, Massachusetts, Minnesota, Oregon, and Utah—also had a PPG and generally agreed that EPA's planning and budget process fostered collaboration in setting joint priorities, roles, and accountability for each party. In addition, Minnesota and Massachusetts officials told us that they have been able to use grant resources to address other state priorities. However, these officials noted that EPA guidance continues to call for a specific number of enforcement activities, which does not allow for as much flexibility as envisioned.

States without a PPA can still participate in NEPPS through a PPG, which allows them to combine individual categorical grant funds into a consolidated grant. Once the funds are consolidated, they lose their category-specific identity and can be used with greater flexibility. We found that states that had a PPG, but not a PPA, had mixed views on NEPPS. For example, according to the New York State Department of Environmental Conservation (NYSDEC), which currently has a performance grant only for water programs, it experimented with a performance agreement in the mid-1990s but dropped the endeavor because of difficulties in drafting an agreement that combined multiple agency divisions. NYSDEC officials said the amount of work involved outweighed the benefits to individual programs. Furthermore, even with a performance grant for water programs, NYSDEC officials told us, EPA still maintains requirements for how the money is used and attempts to steer the state toward EPA's priorities, which are not necessarily NYSDEC's priorities. In contrast, the Arizona Department of Environmental Quality, which has a performance grant but not a performance agreement, said it has a "fantastic relationship" with Region 9 that is characterized by extensive coordination and communication. Because of its grant, Arizona officials told us, the state has realized administrative efficiencies and has more flexibility to move money among programs.

States that do not participate in NEPPS were generally satisfied with the amount of joint planning and coordination in their work plan agreements (i.e., memorandum of agreement and annual work plans). For example, West Virginia and Florida told us that their work plan agreements with their EPA regions provided them with much of the same opportunities for joint planning, flexibilities, and priority setting as NEPPS participants. Officials from the West Virginia Department of Environmental Protection said they did not participate in NEPPS because they did not wish to transfer grant funding among media programs. They said state programs rely heavily on revenue generated through fees and penalties; therefore, program managers are reluctant to share resources. According to officials from the Florida Department of Environmental Protection, they have not participated in NEPPS since 1999 because their work plan agreements provide the benefits of NEPPS—good dialogue with the region in planning and priority setting. Moreover, Florida officials said they cannot use the flexibility allowed under NEPPS to redirect resources between programs because state appropriation procedures place restrictions on their ability to shift resources among programs.

STATE REVIEW FRAMEWORK HAS THE POTENTIAL TO PROVIDE MORE CONSISTENT OVERSIGHT

With its implementation of the SRF, EPA has—for the first time—a consistent approach for overseeing authorized states' compliance and enforcement programs and has identified several significant weaknesses in how states enforce their environmental laws in accordance with federal requirements.[29] For example, the SRF reviews found that states are not properly documenting inspection findings or penalties, as directed by EPA's enforcement policy and guidance. While recognizing that these findings are useful, EPA has not developed a plan for how it will uniformly address them in a timely manner. Nor has the agency identified the root causes of the weaknesses, although some EPA and state officials attribute the weaknesses to causes such as increased workloads concomitant with budgetary reductions. Until EPA addresses enforcement weaknesses and their causes, it faces limitations in determining whether states perform timely and appropriate enforcement, and whether they apply penalties to environmental violators in a fair and consistent manner within and among the states. Moreover, the SRF is still in its early stages of implementation and offers additional uses that EPA has not yet considered. In this regard, its structured approach provides consistent information that would be useful to (1) inform the public about how well the states are implementing their enforcement responsibilities and (2) serve as a basis for assessing the performance of EPA's regions, which have been inconsistent in their enforcement and oversight efforts in the past.

SRF Findings Demonstrate the Value of a Uniform Approach to Evaluating State Enforcement Programs

As of January 25, 2007, OECA had conducted SRF reviews in 33 states and expected to complete its assessment of remaining states by the end of fiscal year 2007. OECA reported good performance in most aspects of state compliance and enforcement programs. However, it also reported that the reviews found several weaknesses in state programs that will require focused attention to correct. EPA officials said the following four weaknesses were the most frequently identified:

- States are not adequately documenting the results of facility inspections in order to determine the significance of violations. EPA policy states that complete and accurate documentary evidence is needed to determine the nature and extent of violations, particularly the presence of significant violations, and to support timely and appropriate enforcement actions. EPA policy also states that a quality program should maintain accurate and up-to-date files and records, and report this information to EPA to support effective program evaluation and priority setting. EPA and state officials suggested that some of the causes of inadequate state documentation and reporting of facility inspections can be traced to a lack of staff expertise, inadequate training, increasing workload, and reductions in staff and budgetary resources in recent years.

- States are not adequately entering significant violations noted in their inspection reports into EPA databases. GAO, EPA's Office of Inspector General, and OECA have reported that the lack of complete and accurate national enforcement data has been a long-term problem. OECA needs accurate and complete enforcement data to help it determine whether core enforcement requirements are being consistently implemented by regions and states and whether there are significant variations from these requirements that should be corrected. In addition, accurate and complete enforcement data helps OECA more efficiently and effectively oversee states and regions. According to EPA regional and state officials, the SRF is helping them recognize the reasons for discrepancies between state and EPA databases and improve the quality of data in these databases.

- States lack adequate or appropriate penalty authority or policies. Penalties play a key role in environmental enforcement by deterring potential violators and ensuring that members of the regulated community cannot gain a competitive advantage by violating enforcement regulations. To qualify as an authorized state for administering environmental programs, states must, among other things, acquire and maintain adequate authority to enforce program requirements consistent with federal requirements. For example, to obtain EPA's approval to administer the Clean Air Act's Title V permitting requirements for major air pollution sources, states must have, among other things, authority to recover civil penalties and provide appropriate criminal penalties. EPA's policy provides that all penalties should include two components. First, penalties should include an "economic benefit" component that reflects the benefit achieved by avoiding compliance. This component is considered important to "leveling the playing field" among companies within an industry and eliminating any economic advantage violators gain through delayed or avoided compliance costs. The second component—the "gravity-based component"—reflects the seriousness of the violation, the actual or possible harm it causes, and the size of the violator.

- States are not documenting how they implement EPA's policies for calculating and assessing penalties. According to EPA policy, states need to maintain sufficient documentation so that the regions can evaluate (1) the state's rationale for obtaining a penalty and (2) where appropriate, the calculation of the economic benefit and gravity-based components. If a state has not assessed a penalty or other appropriate sanction against a violator, EPA may take direct enforcement action to recover a penalty. However, regional and state offices we interviewed said applying this policy places strains on the EPA/state working relationship because states generally prefer that EPA not take direct enforcement action against regulated entities in their states. For this reason, EPA generally does not take direct enforcement action solely to

Environmental Protection

recover additional penalties unless a state penalty is determined to be grossly deficient.

Regional and state officials said states vary as to whether their environmental program administrators have the authority to assess penalties. If program administrators lack this authority, they must pursue judicial remedies through their state attorney general. Judicial actions generally result in penalties and court orders requiring correction of the violation. However, this route is more time-consuming and resource- intensive than having the environmental program office assess the penalty and can significantly delay obtaining penalties and achieving a return to compliance.

Modifications to the SRF Could Improve EPA's Enforcement Oversight Process

The initial SRF findings provide the basis for discussions between EPA and the states on how to address deficiencies in state compliance and enforcement programs, many of which have been well known and longstanding. According to OECA officials, EPA will have completed an SRF review in all states by the end of 2007 and will perform an evaluation of the SRF in fiscal year 2008. The proposed evaluation methodology states that EPA will address the effectiveness of implementing the SRF by surveying the state environmental agencies that participated in the reviews and the pollution control associations. However, this proposed evaluation may not yield the results EPA will need to better ensure more consistent state performance and more consistent state program oversight. In this regard, the proposed evaluation methodology does not specify how EPA plans to use the results of the SRF to begin determining the causes for cited deficiencies and to identify strategies for uniformly and expeditiously implementing potential corrective actions.

While EPA's proposed evaluation may prove to be useful in obtaining views on the success of the SRF and on areas needing improvement in performing the reviews, the proposal is still in its early stages of development, and specific details have not been laid out on whether and how certain potential issues will be addressed. EPA could better ensure more consistent state performance and more consistent oversight of state enforcement programs by (1) developing and implementing corrective actions for the major deficiencies identified through the SRF and (2) assessing the capacity of poorly performing state programs to determine whether they possess the staff, financial, and other resources to effectively implement enforcement programs consistent with federal requirements.

Region 8's experience in using the findings of its Uniform Enforcement Oversight System may prove to be useful for OECA in addressing the findings of the SRF on a nationwide basis. This oversight system was designated a "best practice" in the area of state agency oversight by EPA's Office of Inspector General.[30] The region evaluated states' enforcement programs using uniform review criteria that had been previously agreed to by each state agency. The region then used the evaluation findings to develop an improvement strategy tailored to the particular weakness identified in each state's enforcement program. In order to hold the states accountable for correcting the weaknesses identified, Region 8 also used the improvement strategy in arriving at annual agreements with the states. For example,

74 United States Government Accountability Office

because of the findings from its oversight review of Colorado's hazardous waste program, Region 8 officials became concerned about the adequacy of the authorized program's ability to protect human health and the environment or to take on new program responsibilities. On the basis of these concerns, the region conducted an in-depth "capability assessment" of Colorado's hazardous waste program beginning in 1998, covering items such as the state's levels of resources and staff skills; professional development and training programs; and the efforts to resolve legal and institutional limitations in its program. Over the next 4 years, regional and state officials worked with state government leaders to increase program funding, staff, and training, and to implement new penalty policies. Region 8 also directly implemented compliance monitoring and enforcement activities at selected Colorado hazardous waste facilities while the state program developed and implemented its enhanced capabilities. Region 8 conducted another capability assessment in 2001 and concluded that the state had made the improvements necessary to implement a fully authorized program.

EPA could also use the SRF to inform the general public and others about the extent to which states effectively implement environmental enforcement programs consistent with federal requirements. EPA has not yet determined whether or how the results of the SRF reports will be made available to the general public, congressional committees, environmental interest groups, state environmental organizations, local community groups, and other interested groups and organizations. Many of these groups have expressed deep interest in and concern over the years about the consistency, fairness, and effectiveness of environmental enforcement. A common criticism has been that variation in environmental compliance and enforcement among the states has resulted in the lack of equitable public health and environmental protection and the lack of a "level playing field" for business from one state to another. EPA's Office of Inspector General also recommended that regional evaluations of state programs should be made easily accessible to the public as an important means for holding states accountable for their environmental performance. ECOS and several states officials we interviewed expressed concern that public dissemination of the SRF reports would be used inappropriately to compare or rank state performance. However, the SRF reports have the potential to convey useful information to both EPA managers and to the public on the extent to which the enforcement program is being implemented consistently and fairly nationwide.

In addition to the usefulness of the SRF in evaluating state enforcement programs, the SRF provides a model that OECA could use to evaluate progress being made by EPA's regions in addressing inconsistencies in enforcement actions and oversight. Although we focused our review on EPA's oversight of state enforcement programs, rather than on the enforcement programs of EPA's 10 regional offices, we testified in June 2006 based on reviews on EPA's enforcement program that the regions vary substantially in the actions they take to enforce environmental requirements, such as the number of inspections performed at regulated facilities and the amount of penalties assessed for noncompliance with environmental regulations.[31] In addition, past EPA Inspector General and OECA evaluations found variations among regions regarding issues such as sufficiently encouraging states to consider economic benefit in calculating penalties, taking more direct federal actions where states were slow to act, and requiring states to report all significant violators.

In our June 2006 testimony, we stated that broad agreement exists among EPA and state enforcement officials on the key factors contributing to variations among regions including (1) differences in philosophy among regional enforcement staff about how best to secure compliance with environmental requirements, (2) differences in state laws and enforcement authorities and the manner in which regions respond to these differences, (3) variations in resources available to both state and regional enforcement offices, (4) the flexibility afforded by EPA policies and guidance that allow states a degree of latitude in their enforcement programs, and (5) incomplete and inadequate enforcement data that hamper EPA's ability to accurately characterize the extent to which variations occur.[32]

Although EPA has noted that some variation in environmental enforcement is necessary to take into account local environmental conditions and concerns, it has acknowledged that similar violations should be met with similar enforcement responses to ensure fair and consistent enforcement and equitable treatment for regulated businesses, regardless of geographic location. Our testimony noted that the SRF was among the initiatives that could make a positive contribution to EPA's efforts to ensure consistent approaches in regional enforcement activities, although it is too early to tell whether the initiative will create a level playing field for the regulated community across the country.

CONCLUSIONS

The SRF initiative provides EPA with a potential means to ensure consistent and effective enforcement among the states, thereby addressing a difficult and long-standing challenge to the agency. EPA's plan to evaluate the SRF in 2008 will provide the agency with an opportunity to obtain information from the regions and the states regarding what does and does not work well in these SRF reviews and to make appropriate corrections to its review methodology. However, the proposed evaluation methodology does not describe how EPA will examine the causes of the significant deficiencies noted during the SRF reviews and develop a strategy for addressing them. If these deficiencies are not addressed in a uniform and timely manner, EPA and the states will not gain the full benefit of the SRF.

Regardless of the extent and effectiveness of oversight reviews to determine the consistency and effectiveness of enforcement programs, corrective actions will not be feasible if states lack sufficient funding, staff levels, expertise, and other resources that are vital to carrying out their enforcement responsibilities. On the basis of the audits it has conducted, the EPA's Office of Inspector General has endorsed the practice of having regions follow up on the deficiencies noted in their reviews of state programs and making their findings public. Likewise, Region 8 demonstrated the value of performing a capacity assessment to understand why deficiencies exist in a state program to demonstrate to decision makers and the public what needs to be improved. Such assessments would provide an improved basis for a truly collaborative approach between the regions and the states during their annual deliberations on partnership agreements and grants in order to address the root causes of problems identified in state enforcement programs.

EPA has not determined whether or how it will share the results of the SRF reviews with the general public and others, including Members of Congress who over the years have raised questions and expressed concerns about the way the enforcement program has been implemented. If Members of Congress and the public fully understand the deficiencies and their significance, then they would be better informed about how to assist EPA and the states in ensuring that public health and the environment are protected and a level playing field is established for regulated facilities.

Although the SRF thus far has been focused on reviewing state enforcement programs, the process could be extended to include the enforcement programs of EPA's regions. As we have previously reported, the regions have long been inconsistent in their oversight of states within their jurisdiction and the enforcement actions they take in order to provide a level playing field for regulated facilities across the nation.

RECOMMENDATIONS FOR EXECUTIVE ACTION

To enhance EPA's oversight of regional and state enforcement activities to implement environmental programs consistent with the requirements of federal statutes and regulations, we recommend that the Administrator of EPA take the following actions:

- include, in EPA's fiscal year 2008 evaluation of the SRF, an assessment of lessons learned and an action plan for determining how significant problems identified in state programs will be uniformly and expeditiously addressed;
- evaluate the capacity of individual authorized state programs, where the SRF finds the state appears to lack sufficient resources (e.g., funding, staff, and expertise), to implement and enforce authorized programs and then develop an action plan to improve that state's capacity;
- publish the SRF findings so that the public will know how well state regulators are enforcing authorized programs and protecting public health and the environmental conditions in their communities; and
- conduct a performance assessment similar to SRF for regional enforcement programs.

John B. Stephenson
Director, Natural Resources and Environment

APPENDIX I: SCOPE AND METHODOLOGY

To assess how the Environmental Protection Agency (EPA) and authorized state agencies work together to deploy resources, plan, set priorities, and define roles and responsibilities for enforcement of and compliance with environmental programs consistent with federal requirements, we (1) identified the federal resources provided to EPA regions and states for enforcement between 1997 and 2006, and obtained EPA regional and states' views on the adequacy of these resources to implement their activities; (2) determined EPA's progress in improving priority setting and enforcement planning with its regions and authorized states;

and (3) examined EPA efforts to improve its oversight of states' enforcement and compliance programs.

For the purpose of this review, we conducted semistructured interviews with officials at the 10 EPA regions and at 10 authorized state agencies. To construct questions for the interviews, we analyzed policies, procedures, and guidance materials EPA has developed and implemented. We synthesized the findings, conclusions, and recommendations contained in reports by us, EPA's Office of Inspector General (OIG), the National Academy of Public Administration (NAPA), and the Environmental Council of the States (ECOS). We also met with officials from EPA Region 4, the Georgia Department of Natural Resources, South Carolina Department of Health, and the New Jersey Department of Environmental Protection to obtain a more thorough understanding of how state agencies work together to plan, set priorities, define roles and responsibilities, and deploy resources for enforcement and compliance of environmental laws.

We used a nonrandom sample of 10 states, which consisted of 1 state from each region:

- Five had a Performance Partnership Agreement (PPA) and Performance Partnership Grant (PPG) with EPA (Iowa, Massachusetts, Minnesota, Oregon, and Utah). These states incorporated an enforcement program into their PPA, along with other essential elements that EPA and state leaders considered important (e.g., jointly agreed priorities, defined roles/responsibilities, and processes for resource deployment).
- Five states did not have a PPA (Arizona, Arkansas, Florida, New York, and West Virginia), but had either a PPG or an alternative working relationship with EPA.33

Our evaluation of 10 selected states cannot be generalized to the other states with authorized programs. However, we met with EPA officials representing all 10 regions, who provided their perspectives about all state programs within their geographic region. In addition, we examined other sources of state involvement, such as information available from ECOS and pollution control associations (e.g., Association of State and Territorial Solid Waste Management Officials, Association of State and Interstate Water Pollution Control Administrators, State and Territorial Air Pollution Program Administrators, and the Association of Local Air Pollution Control Officials, now known as the National Association of Clean Air Agencies).

We also limited our review to the environmental agencies within each selected state that implement the major EPA programs (Clean Water Act, Clean Air Act, Safe Drinking Water Act, and Resource Conservation and Recovery Act) and did not include state departments or agencies that implemented other programs, such as the Federal Insecticide, Fungicide, and Rodenticide Act.

To determine if there were any trends in the federal resources provided to EPA regions and states for enforcement from 1997 to 2006, and assess EPA regional and state views on the adequacy of these resources to implement their activities, we reviewed the budgets for EPA and the Office of Enforcement and Compliance Assurance (OECA) for fiscal years 1997 through 2006. We also reviewed our prior reports and those from the Congressional Research Service (CRS), Congressional Budget Office (CBO), Office of Management and Budget (OMB), EPA, and the EPA OIG, for information on the distribution of federal resources. We met with officials from the EPA's Office of the Chief Financial Officer and OECA, ECOS, and the Natural Resources Defense Council, and administered structured interviews to officials in all 10 EPA regions and the selected state in each region. In each region and state, we obtained perspectives on the deployment of resources. However, we were not able to assess the workload of the states and regions and their overall capability to meet federal enforcement requirements because EPA's data collection system does not collect sufficient information needed to make such an assessment. In this regard, EPA lacks information on the capacity of both the states and EPA's regions to effectively carry out their enforcement programs, because the agency has done little to assess the overall enforcement workload of the states and regions and the number and skills of people needed to implement enforcement tasks, duties, and responsibilities. Furthermore, the states' capacity continues to evolve as they assume a greater role in the day-to-day management of enforcement activities, workload changes occur as a result of new environmental legislation, new technologies are introduced, and state populations shift.

To determine EPA's progress in improving priority setting and enforcement planning with its regions and authorized states, we reviewed EPA's strategic plans and national strategy, its policy and guidance for planning and implementing its enforcement programs, the process for implementing National Environmental Performance Partnership agreements with authorized state agencies, and *Federal Register* notices. We also reviewed our prior reports and those from CRS, OMB, EPA, EPA OIG, and NAPA for information on the planning and priority setting process. We met with officials from EPA's Office of Congressional and Intergovernmental Relations, OECA, ECOS, and the Natural Resources Defense Council, and administered structured interviews to all 10 EPA regions and the selected state in each region. In each region and state, we examined regional and state strategic plans and state-EPA enforcement agreements, such as memorandums of agreement, PPAs, and PPGs. At state agencies, we also discussed the states' perspectives on how EPA administered state-EPA agreements, regional plans, and national priorities.

To determine EPA efforts to improve its oversight of state enforcement and compliance programs, we reviewed EPA's policy and guidance for overseeing state agencies. We met with officials from EPA's Office of Congressional and Intergovernmental Relations, OECA, ECOS, and the Natural Resources Defense Council, and administered structured interviews to all 10 EPA regions, and the selected state in each region. In each region, we examined strategic plans, state-EPA agreements, and 33 state oversight reviews of the SRF. At state agencies, we reviewed policy and guidance and received perspectives on EPA's oversight process.

We performed our work from October 2005 through July 2007, in accordance with generally accepted government auditing standards, which included an assessment of data reliability and internal controls.

Appendix II: Development and Implementation of the State Review Framework

EPA's criteria for assessing the performance of compliance assurance and enforcement responsibilities in authorized states can be traced back to the mid-1980s. In August 1986, the EPA Deputy Administrator issued a policy guidance memorandum entitled "Revised Policy Framework for State/EPA Enforcement Agreements." The policy memorandum was intended to provide a framework for gathering information and making judgments about the effectiveness of state compliance and enforcement performance and providing guidance on when and how EPA would become involved in enforcement actions in authorized states. Among other things, the 1986 policy guidance discussed (1) EPA oversight criteria and the measures that the agency would use to define good state performance, (2) oversight procedures and protocols, and (3) criteria for direct federal intervention— factors EPA would consider before taking direct enforcement action in a state and what states might reasonably expect of EPA in this regard.

According to OECA officials, subsequent experience with the 1986 policy guidance revealed several implementation shortcomings that limited EPA's ability to adequately and consistently oversee state compliance and enforcement programs. For example, the 1986 policy guidance did not clearly describe how EPA would oversee state enforcement programs, including what constituted "good program performance". Moreover, EPA and authorized states did not agree on what uniform program information states needed to maintain and provide to EPA for performance measurement. This led to considerable inconsistency from region to region in overseeing state compliance and enforcement programs.

By 2003, OECA officials said that environmental commissioners in several states and members of ECOS were in the forefront of the call for developing and implementing a more uniform and systematic process for EPA's oversight and evaluation of state compliance and enforcement programs. Other factors also pointed to the need to develop a more consistent method of gauging state performance. These included EPA's OIG audits of state programs, EPA's internal assessments, and public petitions for withdrawal of program authorizations for some state programs. A common criticism was that variation in environmental compliance and enforcement among the states was directly attributable to the lack of uniform EPA oversight and performance measurement and that the result was the lack of equitable public health and environmental protection and lack of a level playing field for business from one state to another.

At an ECOS meeting in 2003, OECA officials said the Chairman of the ECOS Compliance Committee proposed to EPA's then Deputy Assistant Administrator for OECA a method for systematically and uniformly assessing state performance. The assessment method, called the SRF, was patterned after a review process originally developed by EPA's Region 8 to assess the performance of state compliance and enforcement programs in that region (encompassing the states of Colorado, Utah, Wyoming, South Dakota, North Dakota, and Montana and 27 sovereign tribal nations). The SRF was formally agreed to by EPA, ECOS, state media organizations, and state environmental agency officials in December 2003.

The 12-point evaluation model used in Region 8, called the Uniform Enforcement Oversight System, became the basis for a review framework for evaluating state compliance and enforcement performance. These 12 required elements for evaluation of state performance include the following:

1. the degree to which a state program has completed the universe of planned inspections (addressing core requirements and federal, state and regional priorities);
2. the degree to which inspection reports and compliance reviews document inspection findings, including accurate descriptions of what was observed to sufficiently identify violations;
3. the degree to which inspection reports are completed in a timely manner, including timely identification of violations;
4. the degree to which significant violations (e.g., significant noncompliance and high priority violations) and supporting information are accurately identified and reported to EPA national databases in a timely manner;
5. the degree to which state enforcement actions include required corrective or complying actions (e.g., injunctive relief) that will return facilities to compliance in a specific time frame;
6. the degree to which a state takes timely and appropriate enforcement actions, in accordance with policy relating to specific media;
7. the degree to which a state includes both gravity and economic benefit calculations for all penalties, appropriately using the economic benefit calculation model (BEN) or similar state model (where in use and consistent with national policy);
8. the degree to which final enforcement actions collect appropriate economic benefit and gravity penalties in accordance with applicable penalty policies;
9. the degree to which enforcement commitments in the PPA, PPG, and/or other written agreements to deliver a product/project at a specified time, if they exist, are met and any products or projects are completed;
10. the degree to which the minimum data requirements are timely;
11. the degree to which the minimum data requirements are accurate; and
12. the degree to which the minimum data requirements are complete, unless otherwise negotiated by the region and state or prescribed by a national initiative.

The SRF also includes a 13th "optional" element that is open for negotiation between regions and states. EPA and ECOS encourage the use of the 13th element to ensure the review takes a measure of the full range of program activities and results. These components can add meaningful input into a state's overall performance and program. Topics could include program areas such as compliance assistance, pollution prevention, innovation, incentive or self-disclosure programs, relationships with state attorneys general, and outcome measures or environmental indicators that go beyond the core program activities covered in elements 1 through 12.

The SRF was also seen by the parties as being consistent with the principles of the National Environmental Performance Partnership System, which provides a mechanism for joint planning and program management that takes advantage of the unique capabilities of each party in addressing pressing environmental problems. Also, the SRF considers commitments negotiated between EPA regions and states contained in PPAs, PPGs, and/or other agreements that may differ from national policy and guidance and evaluates state performance in terms of those commitments. In cases where states and regions have negotiated different state commitments (e.g., number of inspections) or other activities, states are held accountable for those commitments, although the SRF reviewers may provide feedback that those commitments need to be increased in the future to fully demonstrate an adequate enforcement program.

After a review is completed, the regions prepare a draft report of the findings and conclusions, jointly discuss with the state how major recommendations will be addressed, and provide the draft report to OECA headquarters. OECA reviews the draft reports and provides comments on the regions' analyses and recommendations. OECA expects regions to incorporate these recommendations into the next round of negotiated agreements, where OECA will track and manage the recommendations to conclusion.

Additional anticipated benefits of applying the SRF's elements in a uniform manner included, among others, (1) more strategic resource allocation, (2) reduction of duplicative work, (3) consistent and predictable baseline oversight across states and regions with agreed-upon thresholds for corrective action, (4) differential oversight of state programs based on performance,[34] (5) a level playing field for states in competition for business, and (6) improved public confidence in federal and state compliance and enforcement programs. In addition, the SRF is viewed as providing a basis to establish a dialogue on performance that will lead to improved program management and environmental results.

OECA pilot-tested the SRF during fiscal year 2004 in at least one state in each of EPA's 10 regions. The states that participated in the pilot were all volunteers. The pilot states included Alaska, Arizona, Colorado, Rhode Island, New Jersey, Maryland, Michigan, Missouri, Nebraska, Oklahoma, and South Carolina. OECA also piloted the SRF in one EPA region, Region 10 (Seattle), to test the approach on EPA's direct implementation of the Resource Conservation and Recovery Act (RCRA) and the Clean Water Act National Pollution Discharge Elimination System (NPDES) program in Alaska. EPA used the SRF to evaluate the enforcement performance of three media programs (both the pilot reviews and subsequent reviews): the Clean Air Act Stationary Sources program, the Clean Water Act NPDES program, and (RCRA) Subtitle C hazardous waste program. For each program, the SRF defined the essential elements and then, in a companion Framework Implementation Guide more fully defined how each element is to be applied and measured.

In February 2005, OECA contracted for an evaluation of the pilot review process to determine whether the SRF provides an accurate assessment of state compliance and enforcement activities. The evaluation sought to obtain answers to questions, such as the following: (1) Are the 12 review elements the right ones? (2) Are the right data metrics being used? (3) What barriers were encountered? The evaluation also addressed such implementation issues as whether the reviews could be streamlined and made more efficient; what barriers were encountered in conducting the reviews and how they could be reduced or eliminated; and the consistency of application across the country, specifically any need for and ways to improve consistency; problems with objectivity; and apparent gaps in capability or expertise that need to be addressed.

The pilots were also evaluated to determine what they indicated about the performance of the states, whether corrective actions or differential oversight agreements get codified in grant agreements, and the best way to summarize and communicate the results of the review. The evaluation used the results of the pilots as well as discussions with key stakeholders to support recommendations aimed at helping OECA improve the SRF before implementing it more broadly. For example, while OECA had allowed regions considerable discretion/flexibility on procedures for file selection, the evaluation identified a number of weaknesses in the file selection process and highlighted potential improvements to OECA's file selection protocol. The consultant developed an improved sampling model that would yield representative file reviews across states and provide a more representative picture of enforcement and compliance assurance across states with varying levels of enforcement activity.

APPENDIX III: COMMENTS FROM THE ENVIRONMENTAL AGENCY

UNITED STATES ENVIRONMENTAL PROTECTION AGENCY
WASHINGTON, D.C. 20460

JUL 1 6 2007

OFFICE OF
ENFORCEMENT AND
COMPLIANCE ASSURANCE

John B. Stephenson
Director, National Resources and Environment
U.S. Government Accountability Office
Room 2T23
441 G Street, N.W.
Washington, D.C. 20548

Dear Mr. Stephenson:

Thank you for the opportunity to provide comments on the draft report, "EPA-State Enforcement Partnership Has Improved, but EPA's Oversight Needs Further Enhancement". The draft report provides a thoughtful discussion on a number of areas central to our implementation of the national enforcement program: 1) our partnership with states in priority setting and implementation of enforcement programs; 2) our efforts to maintain a level of consistency across state programs and our Regional offices; and 3) our work to identify impediments to performance and devise resource strategies targeted at those areas.

I am particularly appreciative of the report's acknowledgement of the substantial progress we have made in terms of priority setting and planning with the states, and our use of the State Review Framework (SRF) to enhance our ability to evaluate and oversee state enforcement programs. In both instances, we have used a collaborative approach, one we will continue to employ as we work to fill remaining data gaps, enhance our relationships with states, and develop innovative resource approaches. The Agency looks forward to a continued dialogue with you and your staff as we embark on these important efforts.

Our detailed comments on the draft report are attached. If you have any questions, please contact me at 564-2440 or your staff may contact Gwendolyn Spriggs in our Immediate Office at 564-2439.

Sincerely,

Lynn Buhl
Catherine R. McCabe
Principal Deputy Assistant Administrator

Attachment

Internet Address (URL) • http://www.epa.gov
Recycled/Recyclable • Printed with Vegetable Oil Based Inks on 100% Postconsumer, Process Chlorine Free Recycled Paper

End Notes

[1] Such injunctive relief includes the authority to order a party that is violating a provision of the law to refrain from further violation and to take action to abate or correct the noncompliance.

[2] "Indian country" includes all land within the limits of an Indian reservation under the jurisdiction of the United States government and all dependent Indian communities within the borders of the United States.

[3] GAO, *Environmental Protection Agency: Protecting Human Health and the Environment Through Improved Management*, GAO/RCED-88-101 (Washington, D. C.: Aug. 16, 1988); GAO, *Environmental Protection: Collaborative EPA-State Effort Needed to Improve New Performance Partnership System*, GAO/RCED-99-171 (Washington, D.C.: June 21, 1999); EPA Office of Inspector General, *EPA Needs Better Integration of the National Environmental Performance Partnership System*, No. 2000-M-000828-000011 (Washington, D.C.: Mar. 31, 2000); National Academy of Public Administration, *Environment.Gov: Transforming Environmental Protection for the 21st Century* (Washington, D.C.: November 2000); GAO, *Environmental Compliance and Enforcement: EPA's Effort to Improve and Make More Consistent Its Compliance and Enforcement Activities*, GAO-06-840T (Washington, D.C.: June 2006).

[4] GAO, *Clean Water Act: Improved Resource Planning Would Help EPA Better Respond to Changing Needs and Fiscal Constraints*, GAO-05-721 (Washington, D.C.: July 22, 2005).

[5] These three programs are the Clean Water Act—National Pollutant Discharge Elimination System; Clean Air Act—Stationary Sources Program; and Resource Conservation and Recovery Act—Subtitle C Hazardous Waste Program.

[6] For purposes of this chapter, state agencies include those of the District of Columbia, Puerto Rico, the U.S. Virgin Islands, and the Pacific Islands.

[7] GAO, *Federal-State Environmental Programs—The State Perspective*, CED-80-106 (Washington, D.C.: Aug. 22, 1980).

[8] GAO/RCED-88-101.

[9] ECOS is the national nonprofit, nonpartisan association that represents state and territorial environmental commissioners.

[10] State pollution control associations are national, nonpartisan professional organizations representing state and local pollution control officials. They include, for example, the Association of State and Territorial Solid Waste Management Officials (ASTSWMO), Association of State and Interstate Water Pollution Control Administrators (ASIWPCA), and the National Association of Clean Air Agencies—formerly known as State and Territorial Air Pollution Program Administrators, and Association of Local Air Pollution Control Officials (STAPPA/ALAPCO).

[11] The review elements are based upon compliance and enforcement policies that have been in place for many years, such as EPA's national enforcement response policies, compliance monitoring policies, civil penalty policies, and similar state policies (where in use and consistent with national policies).

[12] When we refer to "real terms," we mean after subtracting out the effect of inflation, i.e., growth in prices. Trends in spending of nominal amounts (also called current dollar or then-year values) may reflect changes in both price and quantity. To evaluate real spending trends, it is necessary to remove the effect of changes in prices. The effect of inflation is removed by deflating the series, a process that requires dividing the nominal value by an appropriate price index. The resulting series can be labeled real, inflation-adjusted, or constant dollars. This chapter used the Gross Domestic Product (GDP) price index to deflate the nominal dollar amounts of budget authority amounts and arrive at inflation adjusted (or real) dollars in 2007 dollars.

[13] The Resource Conservation and Recovery Act of 1976, Pub. L. No. 94-580, 90 Stat. 2795, amends the Solid Waste Disposal Act, Pub. L. No. 89-272, tit. II, 79 Stat. 997 (1965), the first federal law regulating solid wastes—a broad category of materials including such materials as garbage from treatment plants and discarded materials resulting from industrial, commercial, or agricultural activities.

[14] Precise information about the overall changes in region and state enforcement workload associated with the implementation of new requirements—such as information on changes in the numbers of regulated entities—is not available. However, we and others have reported over the years on the impact of changes in the federal environmental programs to the requirements for EPA's and authorized states' enforcement programs. GAO-05-721; Congressional Research Services, RL30798, *Environmental Laws: Summaries of Statutes Administered by the Environmental Protection Agency* (updated Mar. 31, 2005).

[15] GAO-06-840T.

[16] Water Quality Act of 1987, Pub. L. No. 100-4, 101 Stat. 7.

[17] Clean Air Act Amendments, Pub. L. No. 101-549, 104 Stat. 2399 (1990).

[18] GAO, *Clean Air Act: EPA Should Improve the Management of Its Air Toxics Program*, GAO-06-669 (Washington, D.C.: June 23, 2006).

[19] Title III of the 1990 amendments directs EPA to impose technology-based standards, or Maximum Achievable Control Technology (MACT) standards, on industry to reduce emissions. These technology-based standards require the maximum degree of reduction in emissions that EPA determines achievable for new and existing

84 United States Government Accountability Office

[20] The agency faces court-ordered deadlines to complete standards for all of the remaining 54 source categories of small stationary sources by June 15, 2009.

sources, taking into consideration the cost of achieving such reduction, health and environmental impacts, and energy requirements.

[21] Under Title V of the Clean Air Act, sources emitting pollutants above certain thresholds are classified as "major sources" and must obtain Title V operating permits. In addition, most major sources must report their aggregate annual emissions to their state air quality agency and pay fees based partly or entirely on their level of emissions. Sources that emit pollutants below major source thresholds are called "minor sources" and do not have to obtain a Title V permit. Some minor sources, called "synthetic minors," have the potential to emit pollutants at major source levels but choose to limit their operations and emit below these thresholds.

[22] GAO, *Air Pollution: EPA Should Improve Oversight of Emissions Reporting by Large Facilities,* GAO-01-46 (Washington, D.C.: April 2001).

[23] GAO-01-46.

[24] The Hazardous and Solid Waste Amendments of 1984, Pub. L. No. 98-616, 98 Stat. 3221.

[25] Pub. L. No. 109-58, 119 Stat. 594.

[26] The act allows EPA to extend the first 3-year period for up to 1 additional year if an authorized state demonstrates that it has insufficient resources to complete all inspections within the first 3-year period.

[27] The Leaking Underground Storage Tank Trust Fund, as established in 1986, provided funds to states specifically for cleaning up contamination from tanks (i.e., releases or leaks). Prior to the Energy Policy Act of 2005, states could not use the money for inspections or enforcement of leak detection and prevention requirements.

[28] Association of State and Territorial Solid Waste Management Officials also received a collective recommendation from the Northeast Waste Management Officials' Association (NEWMOA), a nonprofit, nonpartisan interstate association that has a membership composed of the hazardous waste, solid waste, waste site cleanup and pollution prevention program directors for the environmental agencies in eight New England States (Connecticut, Maine, Massachusetts, New Hampshire, New Jersey, New York, Rhode Island, and Vermont)

[29] As we reported in June 2006, EPA has long experienced difficulties in providing oversight of state enforcement programs with sufficient consistency.

[30] EPA Office of Inspector General, *Water Enforcement: State Enforcement of Clean Water Act Dischargers Can Be More Effective,* Audit Report No. 2001-P-00013 (Washington, D.C.: Aug. 14, 2001).

[31] GAO-06-840T.

[32] GAO/RCED-00-108.

[33] For the purposes of this chapter, state agencies include those of the District of Columbia, Puerto Rico, and the U.S. Virgin Islands, and the Pacific Islands.

[34] The term "differential oversight" refers to a mechanism through which the compliance and enforcement program can offer differing levels of oversight based on EPA's assessment of state performance. States demonstrating an adequate core compliance and enforcement program would qualify for benefits while state performance not meeting minimum standards would result in enhanced oversight. This process does not negate EPA's responsibility for oversight; it simply determines the level, intensity, and focus of the oversight.

In: Enforcing Federal Pollution Control Laws
Editor: Norbert Forgács

ISBN: 978-1-60876-082-4
© 2010 Nova Science Publishers, Inc.

Chapter 4

ENVIRONMENTAL COMPLIANCE AND ENFORCEMENT: EPA'S EFFORT TO IMPROVE AND MAKE MORE CONSISTENT ITS COMPLIANCE AND ENFORCEMENT ACTIVITIES*

John B. Stephenson

WHY GAO DID THIS STUDY

The Environmental Protection Agency (EPA) enforces the nation's environmental laws and regulations through its Office of Enforcement and Compliance Assurance (OECA). While OECA provides overall direction on enforcement policies and occasionally takes direct enforcement action, many enforcement responsibilities are carried out by EPA's 10 regional offices. In addition, these offices oversee the enforcement programs of state agencies that have been delegated the authority to enforce federal environmental protection regulations.

This testimony is based on GAO's reports on EPA's enforcement activities issued over the past several years and on observations from ongoing work that is being performed at the request of this Committee and the Subcommittee on Interior, Environment and Related Agencies, House Committee on Appropriations. GAO's previous reports examined the (1) consistency among EPA regions in carrying out enforcement activities, (2) factors that contribute to any inconsistency, and (3) EPA's actions to address these factors. Our current work examines how EPA, in consultation with regions and states, sets priorities for compliance and enforcement and how the agency and states determine respective compliance and enforcement roles and responsibilities and allocate resources for these purposes.

---

* This is an edited, reformatted and augmented version of a U. S. Government Accountability Office publication, Report GAO-06-840-T, dated June 2006.

WHAT GAO FOUND

EPA regions vary substantially in the actions they take to enforce environmental requirements, according to GAO's analysis of key management indicators that EPA headquarters uses to monitor regional performance. These indicators include the number of inspections performed at regulated facilities and the amount of penalties assessed for noncompliance with environmental regulations. In addition, the regions differ substantially in their overall strategies to oversee states within their jurisdictions. For example, contrary to EPA policy, some regions did not require states to report all significant violators, while other regions adhered to EPA's policy in this regard.

GAO identified several factors that contribute to regional variations in enforcement. These factors include (1) differences in philosophy among regional enforcement staff about how best to secure compliance with environmental requirements; (2) incomplete and unreliable enforcement data that impede EPA's ability to accurately determine the extent to which variations occur; and (3) an antiquated workforce planning and allocation system that is not adequate for deploying staff in a manner to ensure consistency and effectiveness in enforcing environmental requirements.

EPA recognizes that while some variation in environmental enforcement is necessary to reflect local conditions, core enforcement requirements must be consistently implemented to ensure fairness and equitable treatment. Consequently, similar violations should be met with similar enforcement responses regardless of geographic location. In response to GAO findings and recommendations, EPA has initiated or planned several long-term actions that are intended to achieve greater consistency in state and regional enforcement actions. These include (1) a new State Review Framework process for measuring states' performance of core enforcement activities, (2) a number of initiatives to improve the agency's compliance and enforcement data, and (3) enhancements to the agency's workforce planning and allocation system to improve the agency's ability to match its staff and technical capabilities with the needs of individual regions. However, these actions have yet to achieve significant results and will likely require a number of years and a steady top-level commitment of staff and financial resources to substantially improve EPA's ability to target enforcement actions in a consistent and equitable manner.

Mr. Chairman and Members of the Committee:

I am pleased to be here today to discuss our work on the Environmental Protection Agency's (EPA) difficulties in ensuring consistent and equitable enforcement actions among its regions and among the states. Our testimony today is based on reports we have issued on EPA's compliance and enforcement activities over the past several years,[1] and provides some observations from the ongoing work that we are performing at your request and that of the Subcommittee on Interior, Environment and Related Agencies, House Committee on Appropriations. As you know, we are assessing how EPA, in consultation with regions and state agencies, sets priorities for compliance and enforcement and how the agency and the states determine respective compliance and enforcement roles and responsibilities and allocate resources for these purposes. As part of this effort, we are assessing EPA's initiated and planned actions to address key factors that result in inconsistencies—identified in our previous work—in carrying out its enforcement responsibilities. We expect to complete this ongoing review on EPA and state enforcement and issue our report in March 2007.

EPA seeks to achieve cleaner air, purer water, and better protected land in many different ways. Compliance with the nation's environmental laws is the goal, and enforcement is a vital part of the effort to encourage state and local governments, companies, and others who are regulated to meet their environmental obligations. Enforcement deters those who might otherwise seek to profit from violating the law, and levels the playing field for environmentally compliant companies.

EPA administers its environmental enforcement responsibilities through its Office of Enforcement and Compliance Assurance (OECA). While OECA provides overall direction on enforcement policies, and occasionally takes direct enforcement action, many of its enforcement responsibilities are carried out by its 10 regional offices (regions). These regions, in addition to taking direct enforcement action, oversee the enforcement programs of state agencies that have been delegated authority for enforcing federal environmental protection requirements.[2]

In my testimony today, I will describe the (1) extent to which variations exist among EPA's regions in enforcing environmental requirements, (2) key factors that contribute to any such variations, and (3) status of the agency's efforts to address these factors.

In summary, as we previously reported on regional efforts to enforce provisions of the Clean Water Act and the Clean Air Act, the regions vary substantially in the actions they take to enforce environmental requirements. These variations show up in key management indicators that EPA headquarters officials have used to monitor regional performance, such as the number of inspections performed at regulated facilities and the amount of penalties assessed for noncompliance with environmental regulations. For example, in fiscal year 2000, the number of inspections conducted under the Clean Air Act compared with the number of facilities in each region subject to EPA's inspection under the act varied from a high of 80 percent in Region 3 to a low of 27 percent in Regions 1 and 2.

We also reported that it is important to understand the reasons for some of these variations, such as a regional determination to conduct more in-depth inspections at a fewer number of facilities instead of conducting less intensive examinations at many more facilities. Accordingly, we recommended that EPA clarify which enforcement actions it expects to see consistently implemented across the regions and direct the regions to supplement its reporting with information that helps explain why variation occurred. We did not focus our work on the effects of inconsistent enforcement on various types of businesses, including small businesses, the particular focus of the Committee's hearing today. However, in performing our work we noted that a recent study for the Small Business Administration,[3] as well as other studies, have suggested that environmental requirements fall most heavily on small businesses. To the extent that this is the case, small businesses could be especially disadvantaged by any inconsistencies and inequities in EPA's enforcement approach. EPA has made progress toward resolving challenges in its enforcement activities that we have previously identified. Nonetheless, each of the challenges is complex and will require much more work and continued vigilance to overcome.

Our work has identified several factors contributing to regional variations: (1) differences in the philosophy of enforcement staff about how to best achieve compliance with environmental requirements; (2) incomplete and inadequate enforcement data, which hamper EPA's ability to accurately determine the extent of variations; and (3) an antiquated workforce planning and allocation system that is not adequate for deploying staff to ensure greater consistency and effectiveness in enforcing environmental requirements.

Finally, EPA recognizes that to ensure fair and equitable treatment, core enforcement requirements must be consistently implemented so that similar violations are met with similar enforcement responses, regardless of geographic location. Accordingly, and in response to our findings and recommendations, the agency has initiated or planned actions that are intended to achieve greater consistency in regional and state enforcement activities. These actions include the following:

- Developing the State Review Framework. This framework involves a new process for conducting reviews and measuring the performance of core enforcement programs in states with delegated authority (as well as nondelegated programs implemented by EPA regions). Although the process is a promising means for ensuring more consistent enforcement actions, it is too early to assess whether the process will result in more consistent enforcement actions and a level playing field for the regulated community across the nation.

-

- Improving management information. EPA has a number of ongoing activities to improve the agency's enforcement data, but the data problems are long-standing and complex. It will likely require a number of years and a steady top-level commitment of staff and financial resources to substantially improve the data so that they can be effectively used to target enforcement actions in a consistent and equitable manner.

-

- Enhancing workforce planning and allocation. For the past several years, EPA has taken measures to improve its ability to match its staff and technical capabilities with the needs of individual regions and states. For example, EPA developed a human capital strategy and performed a study of its workforce competencies. Nonetheless, the agency still needs to determine how to deploy its employees among its strategic goals and geographic locations so that it can most effectively use its resources, including its compliance and enforcement resources.

REGIONAL ENFORCEMENT ACTIVITIES VARY SUBSTANTIALLY

EPA's enforcement program depends heavily upon inspections by regional or state enforcement staff as the primary means of detecting violations and evaluating overall facility compliance. Thus, the quality and the content of the agency's and states' inspections, and the number of inspections undertaken to ensure adequate coverage, are important indicators of the enforcement program's effectiveness. However, as we reported in 2000, EPA's regional offices varied substantially on the actions they take to enforce the Clean Water Act and Clean Air Act. Consistent with earlier observations of EPA's Office of Inspector General and internal agency studies, we found these variations in regional actions reflected in the (1) number of inspections EPA and state enforcement personnel conducted at facilities discharging pollutants within a region, (2) number and type of enforcement actions taken, and

Environmental Compliance and Enforcement   89

(3) the size of the penalties assessed and the criteria used in determining the penalties assessed. For example, as figure 1 indicates, the number of inspections conducted under the Clean Air Act in fiscal year 2000 compared with the number of facilities in each region subject to EPA's inspection under the act varied from a high of 80 percent in Region 3 to a low of 27 percent in Regions 1 and 2.

While the variations in enforcement raise questions about the need for greater consistency, it is also important to get behind the data to understand the cause of the variations and the extent to which they reflect a problem. For example, EPA attributed the low number of inspections by its Region 5, in Chicago, to the regional office's decision at the time to focus limited resources on performing detailed and resource-intensive investigations of the region's numerous electric power plants, rather than conducting a greater number of less intensive inspections.

We agree that regional data can be easily misinterpreted without the contextual information needed to clarify whether variation in a given instance is inappropriate or whether it reflects the appropriate exercise of flexibility by regions and states to tailor their priorities to their individual needs and circumstances. In this regard, we recommended that it would be appropriate for EPA to (1) clarify which aspects of the enforcement program it expects to see implemented consistently from region to region and which aspects may appropriately be subject to greater variation and (2) supplement region-by-region data with contextual information that helps to explain why variations occur and thereby clarify the extent to which variations are problematic.

Our findings were also consistent with the findings of EPA's Inspector General and OECA that regions vary in the way they oversee state-delegated programs. In this regard, contrary to EPA policy, some regions did not (1) conduct an adequate number of oversight inspections of state programs, (2) sufficiently encourage states to consider economic benefit in calculating penalties, (3) take more direct federal actions where states were slow to act, and (4) require states to report all significant violators. Regional and state officials generally indicated that it was difficult for them to ascertain the extent of variation in regional enforcement activities, given their focus on activities within their own geographic environment. However, EPA headquarters officials responsible for the air and water programs noted that such variation is fairly commonplace and does pose problems. The director of OECA's water enforcement division, for example, told us that, in reacting to similar violations, enforcement responses in certain regions are stronger than they are in others and that such inconsistencies have increased.

Similarly, the director of OECA's air enforcement division said that, given the considerable autonomy of the regional offices, it is not surprising that variations exist in how they approach enforcement and state oversight. In this regard, the director noted, disparities exist among regions in the number and quality of inspections conducted and in the number of permits written in relation to the number of sources requiring permits.

In response to these findings, a number of regions have begun to develop and implement state audit protocols, believing that having such protocols could help them review the state programs within their jurisdiction with greater consistency. Here, too, regional approaches differ. For example:

- Region 1, in Boston, has adopted a comprehensive "multimedia" approach in which it simultaneously audits all of a state's delegated environmental programs.

- Region 3, in Philadelphia, favors a more targeted approach in which air, water, and waste programs are audited individually.
- In Region 5, in Chicago, the office's air enforcement branch chief said that he did not view an audit protocol as particularly useful, noting that he prefers regional staff to engage in joint inspections with states to assess the states' performance in the field and to take direct federal action when a state action is inadequate.

We recognize the potential of these protocols to achieve greater consistency by a region in its oversight of its states, and the need to tailor such protocols to meet regional concerns. However, we also believe that EPA guidance on key elements that should be common to all protocols would help engender a higher level of consistency among all 10 regions in how they oversee states.

SEVERAL FACTORS CONTRIBUTE TO VARIATIONS IN REGIONAL ENFORCEMENT PROGRAMS

While EPA's data show variations in key measures associated with the agency's enforcement program, they do little to explain the causes of the variations. Without information on causes, it is difficult to determine the extent to which variations represent a problem, are preventable, or reflect appropriate regional and state flexibility in applying national program goals to unique circumstances. Our work identified the following causes: (1) differences in philosophical approaches to enforcement, (2) incomplete and inaccurate national enforcement data, and (3) an antiquated workforce planning and allocation system.

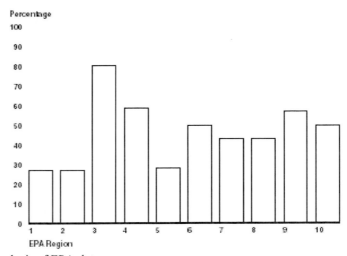

Source: GAO's analysis of EPA data.

Figure 1. Percentage of Total Regulated Facilities Inspected Under the Clean Air Act During Fiscal Year 2000, by EPA Region.

Regions Differ in Their Philosophical Approaches to Enforcement

While OECA has issued policies, memorandums, and other documents to guide regions in their approach to enforcement, the considerable autonomy built into EPA's decentralized, multilevel organizational structure allows regional offices considerable latitude in adapting headquarters' direction in a way they believe best suits their jurisdiction. The variations we identified often reflect different enforcement approaches in determining whether the region should (1) rely predominantly on fines and other traditional enforcement methods to deter noncompliance and to bring violators into compliance or (2) place greater reliance on alternative strategies, such as compliance assistance (workshops, site visits, and other activities to identify and resolve potential compliance problems). Regions have also differed on whether deterrence could be achieved best through a small number of high-profile, resource-intensive cases or a larger number of smaller cases that establish a more widespread, albeit lower profile, enforcement presence. Further complicating matters are the wide differences among states in their enforcement approaches and the various ways in which regions respond to these differences. Some regions step more readily into cases when they consider a state's action to be inadequate, while other regions are more concerned about infringing on the discretion of states that have been delegated enforcement responsibilities. While all of these approaches may be permissible, EPA has experienced problems in identifying and communicating the extent to which variation either represents a problem or the appropriate exercise of flexibility by regions and states to apply national program goals to their unique circumstances.

National Enforcement Data are Incomplete and Inaccurate

OECA needs accurate and complete enforcement data to determine whether regions and states are consistently implementing core program requirements and, if not, whether significant variations in meeting these requirements should be corrected. The region or the state responsible for carrying out the enforcement program is responsible for entering data into EPA's national databases. However, both the quality of and quality controls over these data were criticized by state and regional staff we interviewed.

Internal OECA studies have also acknowledged the seriousness of the data problem. An OECA work group, the "Targeting Program Review Team," stated that key functions related to data quality, such as the consistent entry of information by regions and states, were not working properly and that there were important information gaps in EPA's enforcement-related databases. Another OECA work group concluded in 2006, "OECA managers do not have available to them timely, complete, and detailed analyses of regional or national performance." A third OECA work group asserted that the situation has deteriorated from past years, noting:

> "managers in the regions and in OECA headquarters have become increasingly frustrated that they are not receiving from [the Office of Compliance] the reports and data analyses they need to manage their programs... [and there] has been less attention to the data in the national systems, a commensurate decline in data quality, and insufficient use of data by enforcement/compliance managers."

Consistent with our findings and recommendations, EPA's Office of Inspector General recently reported that, "OECA's 2005 publicly-reported GPRA [Government Performance and Results Act] performance measures do not effectively characterize changes in compliance or other outcomes because OECA lacks reliable compliance rates and other reliable outcome data. In the absence of compliance rates, OECA reports proxies for compliance to the public and does not know if compliance is actually going up or down. As a result, OECA does not have all the data it needs to make management and program decisions. What is missing most, the biggest gap, is information about compliance rates. OECA cannot demonstrate the reliability of other measures because it has not verified that estimated, predicted, or facility self-reported outcomes actually took place. Some measures do not clearly link to OECA's strategic goals. Finally, OECA frequently changed its performance measures from year to year, which reduced transparency." For example, between fiscal years 1999-2005, OECA reported on a low of 23 performance measures to a high of 69 measures, depending on the fiscal year.

Although EPA is working to improve its data, the problems are extensive and complex. For example, the Inspector General recently reported that OECA cannot generate programmatic compliance information for five of six program areas; lacks knowledge of the number, location, and levels of compliance for a significant portion of its regulated universe; and concentrates most of its regulatory activities on large entities and knows little about the identities or cumulative impact of small entities. Consequently, the Inspector General reported, OECA currently cannot develop programmatic compliance information, adequately report on the size of the universe for which it maintains responsibility, or rely on the regulated universe data to assess the effectiveness of enforcement strategies.[4]

EPA's Workforce Planning and Allocation System Is Not Adequate for Effectively Deploying Staff to Regions

As we reported, EPA's process for budgeting and allocating resources does not fully consider the agency's current workload, either for specific statutory requirements, such as those included in the Clean Water Act, or for broader goals and objectives in the agency's strategic plan. Instead, in preparing its requests for funding and staffing, EPA makes incremental adjustments, largely based on historical precedents, and thus its process does not reflect a bottom-up review of the nature or distribution of the current workload. While EPA has initiated several projects over the past decade to improve its workload and workforce assessment systems, it continues to face major challenges in this area

If EPA is to substantially improve its resource planning, we reported, it must adopt a more rigorous and systematic process for (1) obtaining reliable data on key workload indicators, such as the quality of water in particular areas, which can be used to budget and allocate resources, and (2) designing budget and cost accounting systems that are able to isolate the resources needed and allocated to key enforcement activities.

Without reliable workforce information, EPA cannot ensure consistency in its enforcement activities by hiring the right number or type of staff or allocating existing staff resources to meet current or future needs. In this regard, since 1990, EPA has hired thousands of employees without systematically considering the workforce impact of changes in environmental statutes and regulations, technological advances in affecting the skills and expertise needed to conduct enforcement actions, or the expansion in state environmental staff. EPA has yet to factor these workforce changes into its allocation of existing staff resources to its headquarters and regional offices to meet its strategic goals. Consequently, should EPA either downsize or increase its enforcement and compliance staff, it would not have the information needed to determine how many employees are appropriate, what technical skills they must have, and how best to allocate employees among strategic goals and geographic locations in order to ensure that reductions or increases could be absorbed with minimal adverse impacts in carrying out the agency's mission.

EPA Has Initiated or Planned Actions to Achieve Greater Consistency in Enforcement Activities

Over the past several years, EPA has initiated or planned several actions to improve its enforcement program. We believe that a few of these actions hold particular promise for addressing inconsistencies in regional enforcement activities. These actions include (1) the creation of a State Review Framework, (2) improvements in the quality of enforcement data, and (3) enhancements to the agency's workforce planning and allocation system.

EPA's State Review Framework Holds Promise, but It Is Too Early to Assess Its Effectiveness

The State Review Framework is a new process for conducting performance reviews of enforcement and compliance activities in the states (as well as for nondelegated programs implemented by EPA regions). These reviews are intended to provide a mechanism by which EPA can ensure a consistent level of environmental and public health protection across the country. OECA is in the second year of a 3-year project to make State Review Framework reviews an integral part of the regional and state oversight and planning process and to integrate any regional or state corrective or follow-up actions into working agreements between headquarters, regions, and states. It is too early to assess whether the process will provide an effective means for ensuring more consistent enforcement actions and oversight of state programs to help ensure a level playing field for the regulated community across the country. Issues that still need to be addressed include how EPA will assess states' implementation of alternative enforcement and compliance strategies, such as strategies to assist businesses in their efforts to comply with environmental regulations; encourage businesses to take steps to reduce pollution; offer incentives (e.g., public recognition) for businesses that demonstrate good records of compliance; and encourage businesses to participate in programs to audit their environmental performance and make the results of these audits and corrective actions available to EPA, other environmental regulators, and the public.

Efforts Are Underway to Improve Data, but Critical Gaps Remain

Regardless of other improvements EPA makes to the enforcement program, it needs to have sufficient environmental data to measure changes in environmental conditions, assess the effectiveness of the program, and make decisions about resource allocations. Through its Environmental Indicators Initiative and other efforts, EPA has made some progress in addressing critical data gaps in the agency's environmental information. However, the agency still has a long way to go in obtaining the data it needs to manage for environmental results and needs to work with its state and other partners to build on its efforts to fill critical gaps in environmental data. Filling such gaps in EPA's knowledge of environmental conditions and trends should, in turn, translate into better approaches in allocating funds to achieve desired environmental results. Such knowledge will be useful in making future decisions related to strategic planning, resource allocations, and program management.

Nevertheless, most of the performance measures that EPA and the states are still using focus on outputs rather than on results, such as the number of environmental pollution permits issued, the number of environmental standards established, and the number of facilities inspected. These types of measures can provide important information for EPA and state managers to use in managing their programs, but they do not reflect the actual environmental outcomes that EPA must know in order to ensure that resources are being allocated in the most cost-effective ways to improve environmental conditions and public health.

EPA also has worked with the states and regional offices to improve enforcement data in its Permit Compliance System and believes that its efforts have improved data quality. EPA officials said that the system will be incorporated into the Integrated Compliance Information System, which is being phased in this year. According to information EPA provided, the modernization effort will identify the data elements to be entered and maintained by the states and regions and will include additional data entry for minor facilities and special regulatory program areas, such as concentrated animal feeding operations, combined sewer overflows, and storm water. Regarding the National Water Quality Inventory, the Office of Water recently began advocating the use of standardized, probability-based, statistical surveys of state waters so that water quality information would be comparable among states and from year-to-year.

While these efforts are steps in the right direction, progress in this area has been slow and the benefits of initiatives currently in the discussion or planning stages are likely to be years away from realization. For example, initiatives to improve EPA's ability to manage for environmental results are essentially long-term. They will require a long-term commitment of management attention, follow-through, and support—including the dedication of appropriate and sufficient resources—for their potential to be fully realized. A number of similar initiatives in the past have been short-lived and unproductive in terms of lasting contributions to improved performance management. The ultimate payoff will depend on how fully EPA's organization and management support these initiatives and the extent to which identified needs are addressed in a determined, systematic, and sustained fashion over the next several years.

EPA Has Improved the Management of its Human Capital System, but Challenges Remain in Allocating Staff to Match Enforcement Requirements in its Regions

Since the late 1990s, EPA has made progress in improving the management of its human capital. EPA's human capital strategic plan was designed to ensure a systematic process for identifying the agency's human capital requirements to meet strategic goals. Furthermore, EPA's strategic planning includes a cross-goal strategy to link strategic planning efforts to the agency's human capital strategy. Despite such progress, effectively implementing a human capital strategic plan remains a major challenge. Consequently, the agency needs to continue monitoring progress in developing a system that will ensure a well-trained and motivated workforce with the right mix of skills and experience. In this regard, the agency still has not taken the actions that we recommended in July 2001 to comprehensively assess its workforce—how many employees it needs to accomplish its mission, what and where technical skills are required, and how best to allocate employees among EPA's strategic goals and geographic locations. Furthermore, as previously mentioned, EPA's process for budgeting and allocating resources does not fully consider the agency's current workload. With prior years' allocations as the baseline, year-to-year changes are marginal and occur in response to (1) direction from the Office of Management and Budget and the Congress, (2) spending caps imposed by EPA's Office of the Chief Financial Officer, and (3) priorities negotiated by senior agency managers.

EPA's program offices and regions have some flexibility in realigning resources based on their actual workload, but the overall impact of these changes is also minor, according to agency officials. Changes at the margin may not be sufficient because both the nature and distribution of the workload have changed as the scope of activities regulated has increased and as EPA has taken on new responsibilities while shifting others to the states. For example, controls over pollution from storm water and animal waste at concentrated feeding operations have increased the number of regulated entities by hundreds of thousands and required more resources in some regions of the country. However, EPA may be unable to respond effectively to changing needs and constrained resources because it does not have a system in place to conduct periodic "bottom-up" assessments of the work that needs to be done, the distribution of the workload, or the staff and other resource needs.

Mr. Chairman, to its credit, EPA has initiated a number of actions to improve its enforcement activities and has invested considerable time and resources to make these activities more effective and efficient. While we applaud EPA's actions, they have thus far achieved only limited success and illustrate both the importance and the difficulty of addressing the long-standing problems in ensuring the consistent application of enforcement requirements, fines and penalties for violations of requirements, and the oversight of state environmental programs. To finish the job, EPA must remain committed to continuing the steps that it has already taken. In this regard, given the difficulties of the improvements that EPA is attempting to make and the time likely to be required to achieve them, it is important that the agency remain vigilant. It needs to guard against any erosion of its efforts by factors that have hampered past efforts to improve its operations, such as changes in top management and priorities and constraints on available resources.

Mr. Chairman, this concludes my prepared statement. I would be happy to respond to any questions that you or Members of the Committee may have.

GAO's MISSION

The Government Accountability Office, the audit, evaluation and investigative arm of Congress, exists to support Congress in meeting its constitutional responsibilities and to help improve the performance and accountability of the federal government for the American people. GAO examines the use of public funds; evaluates federal programs and policies; and provides analyses, recommendations, and other assistance to help Congress make informed oversight, policy, and funding decisions. GAO's commitment to good government is reflected in its core values of accountability, integrity, and reliability.

End Notes

[1] See GAO, *Environmental Protection: More Consistency Needed Among EPA Regions in Approach to Enforcement*, GAO/RCED-00-108 (Washington, D.C.: June 2, 2000); *Human Capital: Implementing an Effective Workforce Strategy Would Help EPA to Achieve Its Strategic Goals*, GAO-01-812 (Washington, D.C.: July 31, 2001); and *Clean Water Act: Improved Resource Planning Would Help EPA Better Respond to Changing Needs and Fiscal Constraints*, GAO-05-721 (Washington, D.C.: July 22, 2005).

[2] For many federal environmental programs, EPA either authorizes states to administer the federal program or retains authority to administer the program for the state. The state programs that have been approved by EPA are described as "delegated" in this testimony for clarity and consistency with EPA program terminology.

[3] W. Mark Crain, *The Impact of Regulatory Costs on Small Firms*, a report prepared at the request of the Small Business Administration's Office of Advocacy (Washington, D.C., September 2005).

[4] EPA Office of Inspector General, Limited Knowledge of the Universe of Regulated Entities Impedes EPA's Ability to Demonstrate Changes in Regulatory Compliance, Report No. 2005-P-00024, September 19, 2005.

In: Enforcing Federal Pollution Control Laws
Editor: Norbert Forgács

ISBN: 978-1-60876-082-4
© 2010 Nova Science Publishers, Inc.

Chapter 5

FEDERAL POLLUTION CONTROL LAWS: HOW ARE THEY ENFORCED?[*]

Robert Esworthy

SUMMARY

As a result of enforcement actions and settlements for noncompliance with federal pollution control requirements, the U.S. Environmental Protection Agency (EPA) reported that, for FY2008, regulated entities committed to invest an estimated $11.8 billion for judicially mandated controls and cleanup, and for implementing mutually agreed upon (supplemental) environmentally beneficial projects. EPA estimates that these efforts achieved commitments to reduce 3.9 billion pounds of pollutants in the environment, primarily from air and water. EPA also assessed more than $195 million in civil and criminal fines and restitution during FY2008. Nevertheless, noncompliance with federal pollution control laws remains a continuing concern. The overall effectiveness of the current enforcement organizational framework, the balance between state autonomy and federal oversight, and the adequacy of funding are long-standing congressional concerns.

This chapter provides an overview of the statutory framework, key players, infrastructure, resources, tools, and operations associated with enforcement and compliance of the major pollution control laws and regulations administered by EPA. It also outlines the roles of federal (including regional offices) and state regulators, as well as the regulated community. Understanding the many facets of how all federal pollution control laws are enforced, and the responsible parties involved, can be challenging. Enforcement of the considerable body of these laws involves a complex framework and organizational setting.

The array of enforcement/compliance tools employed to achieve and maintain compliance includes monitoring, investigation, administrative and judicial (civil and criminal) actions and penalties, and compliance assistance and incentive approaches. Most compliance

---

[*] This is an edited, reformatted and augmented version of a CRS Report for Congress publication, Report RL34384, dated January 2009.

violations are resolved administratively by the states and EPA. EPA concluded 2,084 final administrative penalty orders in FY2008. Civil judicial actions, which may be filed by states or EPA, are the next most frequent enforcement action. EPA may refer civil cases to the Department of Justice (DOJ), referring 280 civil cases in FY2008. The U.S. Attorney General's Office and DOJ's Environmental Crimes Section, or the State Attorneys General, in coordination with EPA criminal investigators and general counsel, may prosecute criminal violations against individuals or entities who knowingly disregard environmental laws or are criminally negligent.

Federal appropriations for environmental enforcement and compliance activities have remained relatively constant in recent fiscal years. Total funding for EPA's enforcement activities in FY2008 was $553.5 million. Many contend that overall funding for enforcement activities has not kept pace with inflation or with the increasingly complex federal pollution control requirements.

INTRODUCTION

Congress has enacted laws requiring individuals and facilities to take measures to protect environmental quality and public health by limiting potentially harmful emissions and discharges, and remediating damage. Enforcement of federal pollution control laws in the United States occurs within a highly diverse, complex, and dynamic statutory framework and organizational setting.

Multiple statutes address a number of environmental pollution issues, such as those associated with air emissions, water discharges, hazardous wastes, and toxic substances in commerce. Regulators and citizens take action to enforce regulatory requirements in a variety of ways to bring violators into compliance, to deter sources from violating the requirements, or to clean up contamination (which may have occurred prior to passage of the statutes). Implementation and enforcement provisions vary substantially from statute to statute, and are often driven by specific circumstances associated with a particular pollution concern. Given these many factors, it is difficult to generalize about environmental enforcement.

This chapter focuses on enforcement of federal environmental pollution control requirements under the Clean Air Act (CAA); the Clean Water Act (CWA); the Comprehensive Environmental Response, Compensation, and Liability Act, (CERCLA or Superfund); and other statutes for which EPA is the primary federal implementing agency.[1] The report provides a brief synopsis of the statutory framework that serves as the basis for pollution control enforcement, including an overview of the key players responsible for correcting violations and maintaining compliance. Implementation and enforcement of pollution control laws are interdependent and carried out by a wide range of actors including federal, state, tribal and local governments; the regulated entities themselves; the courts; interest groups; and the general public. **Figure 1**, below, presents the array of local, state, tribal, and federal entities that constitutes the environmental pollution control enforcement/compliance framework and organizational setting.

A diverse set of regulatory approaches and enforcement tools are applied to a sizeable universe of regulated entities by these multiple regulating authorities to ensure compliance. A general discussion of enforcement monitoring and response tools is included in this chapter, followed by a summary of recent federal funding levels for enforcement activities. Discussion of available enforcement data sources, as well as tables illustrating examples of trends in enforcement activities, is presented in the two appendices.

While this chapter touches on many aspects of environmental enforcement, it does not describe every aspect and statute in detail. Rather, the report is intended to provide a broad perspective of environmental enforcement by highlighting key elements, and a general context for the range of related issues frequently debated. Information included in this chapter is derived from a variety of sources. These sources, including relevant subject-matter CRS reports providing in-depth discussion of specific topics and laws, are referenced throughout.

Several themes reflecting congressional concerns over time since EPA was established in 1970 are reflected throughout the major sections of this chapter. Congress has conducted oversight, primarily in the form of hearings, on various aspects of the organizational infrastructure and operations designed to enforce pollution control statutes. These aspects of enforcement have also been the topic of investigations by the Government Accountability Office (GAO) and EPA's Office of Inspector General (EPA-OIG).[2] The federal government's oversight of and coordination with states in implementing and enforcing federal pollution control laws have been of particular interest to Congress.[3] The following sections briefly discuss some of the key issue areas.

Federal and State Government Interaction

Since many, but not all, of the federal pollution control statutes authorize a substantial role for states, state autonomy versus the extent of federal oversight is often at the center of debate with regard to environmental enforcement. Not unexpectedly, given the "cooperative federalism"[4] that is often used to characterize the federal, state, and tribal governments in the joint implementationc and enforcement of pollution control requirements, relationships and interactions among these key enforcement players often have been less than harmonious.

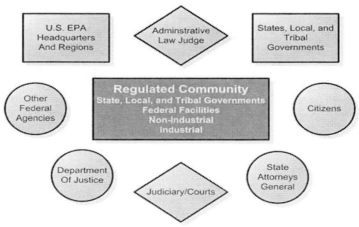

Source: Diagram prepared by CRS.

Figure 1. Key Players in Enforcement of Pollution Control Laws

Disagreements involving environmental priorities and strategic approaches, and balancing the relative roles of compliance assistance with enforcement, contribute to the complexity and friction that come with enforcing national pollution control laws. Other contributing factors include the increasing number of statutory and related regulatory pollution control requirements (some with conflicting mandates) and the adequacy of the resources available for their implementation.

The effects of variability among statutes, coupled with variability in federal and state interpretations and regulations, are often central to the debate. Some argue that this variability leads to too much inconsistency in enforcement actions from state to state, region to region, or between federal versus state actions. Others counter that this represents the flexibility and discretion intended by the statutes to address specific circumstances and pollution problems.

A July 2007 GAO report found that progress had been made regarding federal oversight of state environmental enforcement programs, and that there had been improvements with regard to cooperative federal-state planning and priority setting. However, the GAO concluded that a greater effort was needed to achieve more consistency and effectiveness.[5]

Federal Funding and Staffing for Enforcement Activities

The level of federal funding allocated to states and tribes to support effective enforcement of federal pollution control laws has also been a long-standing congressional concern.[6] In 2004, the Environmental Council of the States (ECOS)[7] reported a $1 billion annual gap in the amount of funding needed by states to implement federal environmental laws, based on a survey of states.[8] GAO reported that, although funding overall for enforcement activities had increased somewhat, it generally had not kept pace with the increasing number of mandates and regulations, or with inflation.[9]

The adequacy of overall federal enforcement funding and personnel, primarily within EPA and the Department of Justice (DOJ), to ensure effective enforcement of environmental statutes, has also been a concern of Congress. Staffing levels of criminal investigators at EPA has been of particular recent interest. For example, Congress enacted the 1990 Pollution Prosecution Act (P.L. 101-593) requiring EPA to hire and maintain 200 criminal investigators. A provision in the House-passed FY2008 Interior and Environmental Agencies Appropriations bill (H.R. 2643) would have required EPA to bring the total number of investigators up to the level of 200 as statutorily required.[10] The provision was not included in the FY2008 appropriations (P.L. 110-161, Title II of Division F). Congressional concerns regarding staff and funding for EPA's criminal (and civil) enforcement were also expressed in conference report language accompanying EPA appropriations for FY2003 through FY2005,[11] and were the topic of a congressionally requested EPA-OIG investigation.[12] Criminal enforcement staffing is discussed further in the "Criminal Judicial Enforcement" section below.

Other Enforcement Issues

Many other aspects of pollution control enforcement have been the subject of debate, and highlighted in congressional hearings and legislation. Some additional areas of continued interest include

- whether there is a need for increased compliance monitoring and reporting by regulated entities;
- impacts of environmental enforcement and associated penalties/fines on federal facilities' budgets (most notably the Department of Defense, or DOD, and Department of Energy, or DOE);
- how best to measure the success and effectiveness of enforcement (e.g., using indicators such as quantified health and environmental benefits versus the number of actions or dollar value of penalties);
- whether penalties are strong enough to serve as a deterrent and maintain a level economic playing field, or too harsh and thus causing undue economic hardship;
- how to balance punishment and deterrence through litigation with compliance assistance, incentive approaches, self-auditing or correction, and voluntary compliance;
- the effect of pollutant trading programs on enforcement; and
- the level of funding required to effectively achieve desired benefits of enforcement.

These issues result from disparate values and perspectives among stakeholders, but also from the factors that are the focus of this chapter: the statutory framework, those who work within this framework, and the tools and approaches that have been adopted for achieving compliance with pollution control laws.

The discussion below, beginning with identification of the principal statutes and key players, followed by an overview of integrated systems of administrative and judicial enforcement, compliance assistance, and incentive tools, is intended to provide a macro-perspective of environmental enforcement infrastructure and operations.

STATUTORY FRAMEWORK FOR ENFORCEMENT OF POLLUTION CONTROL LAWS AND KEY PLAYERS

As Congress has enacted a number of environmental laws over time, as well as major amendments to these statutes, responsibilities of both the regulators and the regulated community have grown. Organizational structures of regulatory agencies have evolved in response to their expanding enforcement obligations. Regulators also must adapt to an evolving, integrated system of administrative and judicial enforcement, compliance assistance, and incentive tools (see discussion under the heading "Enforcement Response and Compliance Tools" later in this chapter).

Table 1. Major Federal Pollution Control Laws

Statute	Major U.S. Code
Comprehensive Environmental Response, Compensation, and Liability Act (Superfund)	42 U.S.C. §§ 9601-9675
Clean Air Act	42 U.S.C. §§ 7401-7671
Clean Water Act	33 U.S.C. §§ 1251-1387
Safe Drinking Water Act	42 U.S.C. §§300f-300j
Solid Waste Disposal Act/Resource Conservation and Recovery Act	42 U.S.C. §§6901-6991k
Environmental Planning and Community-Right-To-Know Act	42 U.S.C. §§11001-110
Federal Insecticide, Fungicide, and Rodenticide Act	7 U.S.C. §§136-136y
Toxic Substances Control Act	15 U.S.C. §2601 et seq.
Pollution Prosecution Act of 1990	42 U.S.C. §4321

Note: This list is not comprehensive in terms of all laws administered by EPA, but rather covers the basic authorities underlying the majority of EPA pollution control programs. For a discussion of these statutes and their provisions, see CRS Report RL30798, *Environmental Laws: Summaries of Major Statutes Administered by the Environmental Protection Agency (EPA)*.

Statutory Framework

The 9 laws listed in **Table 1** generally form the legal basis for the establishment and enforcement of federal pollution control requirements intended to protect human health and the environment.

The discussion in this chapter focuses on these federal environmental laws for which the U.S. Environmental Protection Agency (EPA) is the primary federal implementing agency. Since the EPA was created in 1970, Congress has legislated a considerable body of law and associated programs to protect human health and the environment from harm caused by pollution. Those federal statutes, intended to address a wide range of environmental issues, authorize a number of actions to enforce statutory and regulatory requirements.

Enforcement of this diverse set of statutes is complicated by the range of requirements, which differ based on the specific environmental problem, the environmental media (e.g., air, water, land) affected, the scientific basis and understanding of public risks, the source(s) of the pollutants, and the availability of control technologies. Regulatory requirements range from health and ecologically based numeric standards, or technology-based performance requirements, to facility-level emission and discharge permit limits. Several of the pollution control laws require regulated entities to obtain permits, which typically specify or prohibit certain activities, or delineate allowable levels of pollutant discharges. These permits are often the principal basis for monitoring, demonstrating, and enforcing compliance. In recent years, an increasing number of administrative initiatives have favored incentive-based regulatory approaches, such as trading of permitted emissions, which can affect the applicability of traditional enforcement approaches.

Regulating authorities establish enforcement response and compliance assistance programs to address the enforcement provisions of particular federal pollution control statutes. These environmental statutes typically authorize administrative, civil judicial, and criminal enforcement actions for violations of statutory provisions. For example, §309 of the CWA, § 113 of the CAA, and § 1414 of the Safe Drinking Water Act (SDWA) cover enforcement provisions.[13] As provisions for specific actions vary from statute to statute, each EPA regulatory program office establishes detailed criteria for determining what sanctions are preferable (and authorized) in response to a given violation. The statutes often provide a level of discretion to regulators for addressing specific circumstances surrounding certain environmental problems or violations of national requirements.

Enforcement of the many provisions of the major environmental laws across a vast and diverse regulated community involves a complex coordinated process between federal (primarily EPA and DOJ), state, tribal, and local governments. Congress provided authority to states for implementing and enforcing many aspects of the federal statutory requirements. Citizens also play a role in ensuring that entities comply with environmental requirements, by reporting violations or filing citizen lawsuits, which are authorized under almost all pollution control laws. The following discussion highlights the roles of these key players.

KEY PLAYERS IN ENVIRONMENTAL ENFORCEMENT AND COMPLIANCE

EPA

Primarily through its program offices (e.g., air, water, solid waste), EPA promulgates national regulations and standards.[14] Other federal agencies (e.g., the Department of the Interior, Army Corp of Engineers) and states, tribes, various stakeholder groups, and citizens may contribute input to EPA at various stages of regulatory development (including required public comment). (States may also establish their own laws based on the national requirements; see the discussion later under the heading "States and Delegated Authority"). EPA (and states) inform the regulated community of their responsibilities and administer permitting, monitoring, and reporting requirements. EPA also provides technical and compliance assistance, and employs a variety of administrative and judicial enforcement tools as authorized by the major environmental laws it administers, as well as incentive approaches, to promote and ensure compliance.

Since the EPA's establishment, the Agency's enforcement organization has been modified a number of times, and continues to evolve.[15] EPA's Office of Enforcement and Compliance Assurance (OECA) at headquarters and in the 10 EPA regional offices sets the general framework for federal enforcement activities in coordination with the agency's program offices, states and tribes, and other federal agencies, particularly DOJ. OECA serves as the central authority for developing and implementing a national compliance and enforcement policy, and coordinating and distributing policies and guidance.

EPA's National Program Managers (NPM) Guidance[16] is the primary strategic planning tool that sets out national enforcement program priorities and coordinates state, regional, and EPA headquarters environmental enforcement/compliance activities. The EPA's 10 regional offices, in cooperation with the states, generally are responsible for a significant portion of the day-to-day federal enforcement activities. The NPM Guidance is developed jointly with the EPA regions and states/tribes, and serves as the basis for the enforcement agreements ("commitments") with the regional offices, identifying overall program directions as well as specific activities, allocation of resources, and expected outcomes. On October 12, 2007, OECA announced EPA's current national enforcement and compliance assurance priorities for the years FY2008 through FY2010.[17]

The EPA National Enforcement Investigations Center (NEIC) provides technical expertise to the agency and states. The center administers an investigative team that assigns investigators to the regional offices as needed.[18] OECA also facilitates EPA's National Enforcement Training Institute (NETI), established under Title II of the 1990 Pollution Prosecution Act (P.L. 101-593). NETI provides a wide spectrum of environmental enforcement training online to international, federal, state, local, and enforcement personnel, including lawyers, inspectors, civil and criminal investigators, and technical experts.[19]

OECA's headquarters personnel conduct investigations and pursue or participate in national enforcement cases, particularly those potentially raising issues of national significance. More often enforcement activities fall to the regional offices. EPA (and the states') enforcement actions often require coordination with other federal agencies, most frequently DOJ.

U.S. Department of Justice (DOJ)

In coordination with EPA, the Department of Justice (DOJ)—at its headquarters and through the U.S. Attorneys' offices around the country—plays an integral role in judicial federal enforcement actions of environmental regulations and statutes. EPA refers cases (including some initiated by states) to DOJ for an initial determination of whether to file a case in federal court. DOJ represents EPA in both civil and criminal actions against alleged violators, maintaining close interaction as needed with EPA, states, and tribes during various stages of litigation. DOJ also defends environmental laws, programs, and regulations, and represents EPA when the agency intervenes in, or is sued under, environmental citizen suits. EPA-OECA referred 280 civil cases to DOJ in FY2008[20] and 168 criminal cases in FY2004 (the last year criminal referrals were reported publicly by EPA).[21] Many of these cases are handled by DOJ's Environment and Natural Resources Division (ENRD).[22] EPA and DOJ work conjunctively with the other federal agencies as cases warrant.

Other Federal Agencies

EPA and DOJ coordinate with a number of other federal agencies, particularly when taking criminal action. Key federal agencies include the Federal Bureau of Investigation (FBI), Department of Transportation (DOT), Department of Homeland Security (DHS, particularly the Coast Guard and U.S. Immigration and Customs Enforcement, or ICE), Fish and Wildlife Service, Army Corps of Engineers, Defense Criminal Investigative Service, National Oceanic and Atmospheric Administration (NOAA), U.S. Internal Revenue Service (IRS), and U.S. Securities and Exchange Commission (SEC). These agencies may provide support directly in response to violations of laws implemented by EPA, or, as is often the case, in circumstances where multiple laws have been violated.

States and "Delegated Authority"

Most federal pollution control statutes, but not all, authorize EPA to delegate to states the authority to implement national requirements.[24] For a state to be authorized, or "delegated," to implement a federal environmental program, it must demonstrate the capability to administer aspects of the program's requirements, including the capacity to enforce those requirements. Delegated authority must be authorized under the individual statute, and states must apply for and receive approval from EPA in order to administer (and enforce) federal environmental programs. While many federal pollution control laws provide authority for states to assume primary enforcement responsibilities, there is significant variability across the various laws, including as to standards states must meet and EPA's authority in determining whether states are authorized or have primacy. In some cases, state primacy is almost automatic.

Some federal pollution control laws limit the authority to a specific provision, while others do not authorize delegation at all. For example, §1413 of the Safe Drinking Water Act (SDWA) authorizes states to assume primary oversight and enforcement responsibility (primacy) for public water systems,[25] and §402 of the Clean Water Act (CWA) authorizes state-delegated responsibilities under that act to issue and enforce discharge permits to industries and municipalities. Under CERCLA (Superfund), states are authorized to participate in the cleanup of waste, from taking part in initial site assessment to selecting and carrying out remedial action, and negotiating with responsible parties. Under FIFRA, states may have primacy for enforcing compliance requirements contained on labels of registered pesticides, but are not granted enforcement authority related to registering pesticides or pesticide establishments. Programs under other laws, such as the Toxic Substances Control Act (TSCA), do not provide authority for state delegation. EPA can also authorize state government officials to conduct inspections for environmental compliance on behalf of the agency, subject to the conditions set by EPA, even if a specific statute does not provide delegation authority. However, there must be authority under the specific statute for authorizing such inspections.[26]

Even if delegation is authorized under a federal statute, states may opt not to seek delegation of a particular environmental program, or they may choose only to implement a select requirement under a federal law. For example, as of November 2007, 45 states had obtained the authority to operate the national permitting program under §402 of the CWA, but EPA had only delegated authority to two states to operate the wetlands permitting program under a separate CWA provision, §404.[27]

A majority of states have been delegated authority to implement and enforce one or more provisions of the federal pollution control laws.[28] Authorized states generally implement the national laws and regulations by enacting their own legislation and issuing permits, which must be at least as stringent as the national standards of compliance established by federal law. States consider and approve environmental permits, monitor and assess environmental noncompliance, provide compliance assistance and information to the regulated community and the public, conduct inspections, and take enforcement actions. Local government authorities also may play a role in permitting and monitoring. For example, EPA has delegated authority to implement §112 of the Clean Air Act (CAA) to at least three county governments. However, local governments generally act within the context of assuring states' requirements. For example, local authorities may incorporate land use and other issues as well as code requirements (fire, construction, building safety, plumbing, etc.) in their consideration of permits. A more detailed discussion of the many facets of local authorities is beyond the scope of this chapter.

A significant proportion of inspections and enforcement actions are conducted by the states. Comparable, comprehensive data from the same or similar sources are not readily available for purposes of directly comparing enforcement activities in states relative to EPA. While EPA routinely reports trends in its major enforcement actions in the annual OECA accomplishments reports and on its website, the agency does not include states' activities. There are a number of limitations with regard to states' information currently retained by EPA in its databases (e.g., not all states report relevant information into the EPA databases, reported data are not provided consistently from state to state, and reporting requirements are variable from statute to statute). EPA is working to enhance and improve enforcement reporting by states. The agency has developed and has been implementing its State Review Framework (SRF) tool introduced in 2004, to improve its oversight of state enforcement programs.[29] Under this SRF tool, EPA representatives visit and evaluate each state's compliance and enforcement program based on specified criteria. Through discussions and reports, EPA provides feedback to each state and based on its review, outlines recommendations for improvement. Full implementation of SRF was initiated by EPA in July 2005 and the Agency reported that reviews of all states and territories were completed in 2007. OECA continues to work with its partners to conduct an evaluation implementing SRF recommendations a conducted subsequent reviews.[30] Nevertheless, there are still perceived differences between states, EPA regions, and EPA headquarters.

In recent years, ECOS[31] has served as a forum to improve coordination and promote joint strategic planning between the states and EPA. In addition to other strategic planning tools, EPA and states established the National Environmental Performance Partnership System (performance partnerships, or NEPPS)[32] in 1995 in an effort to improve the effectiveness of EPA-state coordinated environmental management. Under this system, which includes elements of compliance and enforcement, EPA and states enter into individual partnerships (performance partnership agreements) to address jointly agreed-upon priorities based on assessments of localized environmental conditions. The partnerships can be broad in scope or comprehensive strategic plans, and often serve as work plans for funding through EPA grants.

Absent delegation, EPA continues to enforce the federal law in the state, although a state can enforce its own environmental laws where not preempted by federal law. Even with delegation, EPA retains the authority and responsibility as determined by each statute to take enforcement measures, generally taking action when there is a violation of an EPA order or consent decree, or when the federal government deems a state to have failed to respond to a major violation in a "timely and appropriate" manner. Additionally, when a noncompliance case involves an emergency or matters of potential national concern, such as significant risk to public health and safety, the federal government will typically intercede. There are cases where states request the federal government to step in, and other cases where the federal government on its own initiative acts on violations that are the subject of state enforcement action or settlement, known as "overfiling." EPA contends that overfiling occurs infrequently and that certain environmental statutory provisions preclude EPA from overfiling. These provisions are not explicit in all the pollution control statutes, and are limited to specific subsections and violations.[33] Although overfiling of states' enforcement actions has occurred under various pollution control statutes, historically, overfiling of Resource Conservation and Recovery Act (RCRA) violations has been the subject of considerable debate and litigation.

States have strongly objected to overfiling, and the utility and extent of overfiling with respect to environmental enforcement has been the subject of considerable litigation, debate, and literature.[34]

> The term "overfiling" applies to situations when federal enforcement actions are filed during or after a state enforcement action against the same entity for violation of a federal statute. Some states and regulated entities use the term more broadly in reference to assertion of federal authority. Overfiling or the threat of overfilling sometimes strains EPA-state relations and cooperation, sometimes implying criticism of a delegated state's effectiveness.

Tribal Governments

EPA and states increasingly have recognized the role of tribal governments in environmental enforcement, where tribes, rather than states, have primary jurisdiction.[35] Indian tribes, as sovereign governments, can establish and enforce environmental programs under their own laws, but must obtain approval from EPA to administer federal environmental programs on their land. As with states, some of the federal statutes authorize tribes,[36] with EPA approval, to assume responsibility for implementing certain federal pollution control programs. To obtain EPA approval, tribes must demonstrate adequate authority and jurisdiction over the activities and lands to be regulated. Where there is no approved tribal program, EPA exercises its federal authority and may undertake direct program implementation. In some instances, particularly when there are criminal violations, EPA may retain a role in compliance and enforcement even when there is an approved tribal program.

In addition to the federal statutes, a tribal government's authority for environmental protection can arise from federal executive orders, treaties, and agreements with the United States and/or state and local governments,[37] some of which explicitly reserve rights pertaining to the environment. When addressing environmental issues within tribal lands, EPA abides by the January 24, 1983, American Indian policy statement,[38] which reaffirmed the government-togovernment relationship of Indian tribes with the United States.[39]

Relatively few tribes have obtained authority for implementing federal pollution control laws, and EPA identified tribal environmental compliance as a national enforcement and compliance priority in its FY2005-FY2007 enforcement strategic plan in an effort to enhance tribal governments' capabilities to implement federal environmental statutes. The primary focus areas in the tribal strategy, which has been retained as a priority in the FY2008-FY2011 enforcement strategic plan, are public drinking water systems, federal pollution control statutes applicable to schools, and unregulated dumping of solid waste.[40]

Citizens

Private individuals play an important role in enforcing certain aspects of federal pollution control laws. Citizen participation, specifically authorized by Congress in many of the federal pollution control statutes, occurs in several ways. Individuals can identify and report violations of the laws, provide comments on settlements that are reached between the federal government and violators of the environmental laws in enforcement cases, and initiate enforcement proceedings directly in response to alleged violations. In addition, individuals may bring actions against the EPA for failing to execute nondiscretionary duties required under federal environmental laws.[41]

To further enhance public participation and reporting of potential environmental violations, EPA-OECA introduced the "National Report a Violation" website in January 2006.[42] The website provides access to OECA's online citizens' tips and complaints form. EPA reported that the number of citizen tips and complaints increased from 1,485 in FY2005 to 3,274 in FY2006. The majority of these FY2006 citizen submissions (2,800) were referred to civil enforcement, and roughly 480 tips were referred to criminal agents according to EPA.[43]

Regulated Community

The size and diversity of the regulated community are vast, spanning numerous industrial and nonindustrial entities, small and large, and their operations. The following discussion provides an overview of the regulated community, and highlights the role and activities of the key regulated entities in the enforcement of the primary pollution control statutes.

The universe of the regulated community as a whole is very large (see discussion below). The majority of those in the regulated community are required to comply with multiple statutes because of the nature of their activities and operations. The regulated community includes a diverse range of entities and operations, including utilities, refineries, manufacturing and processing facilities, agriculture producers and processors, mobile sources (e.g., private and commercial vehicles), and others. Local, state, tribal and federal governments are also part of the regulated community, as they are engaged in a range of activities and operations—utilities, construction, waste and wastewater management, drinking water management, transportation, pest management—that generate pollution similar to nongovernment sectors.

Regulated entities vary in their activities and operations, and in size—ranging from small individual business operations such as dry-cleaners to facilities and operations that are part of large corporations and conglomerates. Regulatory agencies generally categorize regulated entities into minor and major emitters/dischargers based on factors such as total earnings, number of employees, production volume, and amount of emissions, for purposes of implementing and enforcing the various statutes. In certain circumstances, some of the pollution control statutes make specific distinctions with regard to major and minor emitters/dischargers. A designation of "major" generally applies to those entities that, because of their size or operations, have the potential to have a significant impact on the environment. Most of the statutes and accompanying regulations include authorities for reducing the stringency, and in some cases providing exemptions from regulatory requirements to minimize their impacts on small businesses and operations.

There is no readily available, current, comprehensive list and description of the complete universe of those who are regulated under all of the major pollution control statutes. EPA has been criticized for not adequately defining the regulated universe, a step that GAO determined to be a critical component necessary to evaluate the effectiveness of enforcement.[44] EPA-OECA compiled data regarding the size of the regulated community in September 2001, and estimated a total universe of more than 41 million.[45] Although cited by EPA subsequently from time to time, most commonly in strategic planning documents, the agency has not updated the estimate.

There are, however, data and information that provide some indications of the size and diversity of this universe—for example, in EPA's primary enforcement and compliance databases (see additional discussion in **Appendix**). EPA's publicly available Enforcement and Compliance History Online (ECHO) contains more than 880,000 unique facility records for compliance with CWA, CAA, and RCRA. These records are primarily based on permitted facilities. Another EPA centrally managed database is the Facility Registry System (FRS), which primarily identifies "facilities, sites or places" subject to federal pollution control requirements; it contains more than 1.5 million facility records. The FRS database is primarily based on permit information for CWA, CAA, and RCRA, but includes information reported regarding CERCLA sites. It does not include information indicating the universe regulated under other statutes. In yet another source, the ECOS indicated that states reported that more than 3 million regulated facilities required state agency oversight for environmental compliance in 2003.[46] The differences in the various sources are an indication of the difficulty involved in accurately and consistently tracking the size of the regulated populations.

EPA's various program offices (e.g., air, water, and waste) maintain and publish information and profiles regarding characterizations of regulated entities and their operations. Generally included are estimates of the types and amounts of emissions and discharges, or wastes being handled. For example, EPA's Office of Air and Radiation (OAR) maintains a national database of air emissions estimates for individual point- or major-source categories.[47] The database contains information on stationary and mobile sources that emit common ("criteria") air pollutants [48] and their precursors, as well as hazardous air pollutants (HAPs).[49] The categories presented in these sources do not reflect 100% of the total number of facilities being regulated.

Table 2. EPA Industry and Government Sectors

Available Sector Notebooks		
Aerospace	Food Processing	Power Generators
Agriculture	Furniture	Printing
Automotive	Healthcare	Pulp/Paper/Lumber
Chemicals	Local Government Operations	Rubber/Plastics
Computers/Electronics	Metals	Shipbuilding and Repair
Construction	Minerals/Mining/Processing	Textiles
Dry Cleaning	Petroleum	Transportation
Federal Facilities	Pharmaceuticals	

Source: Table generated by CRS with information from EPA's Sector Compliance Assistance and Sector Notebooks website, http://www.epa.gov/compliance/assistance/sectors/index.html.

Another source for characterizing the sectors of the regulated community are EPA's "Sector Notebooks."[50] EPA has defined sectors as distinct parts of the economy that share similar operations, processes or practices, environmental problems, and compliance issues. EPA recognizes that there are likely a number of circumstances where regulated entities within specific geographic regions may have unique characteristics that are not fully reflected in the profiles contained in the sector notebooks. In addition, some of the notebooks were completed several years ago. Nevertheless, notebook profiles provide fairly comprehensive characterizations of key sectors included within the regulated community. **Table 2** lists the sectors for which the agency has completed sector notebooks.

Enforcement at Federal Facilities

Unless a statutory exemption exists, federal facilities are subject to the federal pollution control statutes,[51] and generally also must adhere to the environmental laws and regulations of the states and municipalities in which they are located, to the same extent as others in the regulated community. EPA reported that it concluded 35 enforcement actions against federal agencies for alleged violations of federal pollution control laws during 2008, resulting in an estimated reduction of more than 1.7 million pounds of pollutants.[52] Federal agencies are also subject to relevant requirements of executive orders.[53]

Regulating federal facilities under pollution control laws presents certain unique challenges. Although all are potentially subject to pollution control laws and regulations, a majority of federal agencies and their facilities are not involved in activities that would generally warrant compliance requirements. According to EPA, facilities operated by DOD and DOE make up a significant portion of the universe of "major" federal facilities.[54] Major federal facilities generally refer to those facilities that, because of their size or operations, have the potential to have a significant impact on the environment. Compliance/enforcement information for DOD and DOE is reported individually, while other federal agencies are generally categorized together as Civilian Federal Agencies.[55] GAO observed that DOD and DOE together accounted for 99% of the federal government's environmental liabilities as of FY2004.[56] Most of these liabilities consist of the estimated cost to safely dispose of solid and hazardous waste generated from day-to-day operations, and to clean contamination from releases into the environment.

The major federal pollution control laws provide EPA with authorities to enforce requirements and impose penalties at federal facilities that are not in compliance. The Federal Facility Compliance Act of 1992 specifically amended RCRA to clarify that DOD and all other federal facilities are subject to penalties, fines, permit fees, reviews of plans or studies, and inspection and monitoring of facilities in connection with federal, state, interstate, or local solid or hazardous waste regulatory programs.[57] The SDWA includes similar language regarding federal facilities, but most of the other federal environmental laws do not include such specific provisions. CERCLA (Superfund) § 120 requires federal agencies with NPL sites to investigate and clean up the contamination, and significantly contaminated federal facility sites have been listed on EPA's National Priorities List (NPL). Whether other pollution control laws should be amended to clarify their applicability to federal facilities has been an issue of debate in Congress.

ENFORCEMENT RESPONSE AND COMPLIANCE TOOLS

EPA and states apply a set of environmental enforcement tools to identify and correct noncompliance, restore environmental damage, and impose penalties intended to deter future violations. Compliance with pollution control laws is addressed through a continuum of response mechanisms, ranging from compliance assistance to administrative and civil enforcement, to the stronger criminal enforcement. The spectra of tools, which escalate in terms of their level of severity and intensity, are authorized in each of the environmental statutes. The following sections of this chapter provide a brief overview of the various enforcement response mechanisms.

Over the years, EPA and states have sought to effectively balance the provision of guidance and assistance to prevent violations or achieve compliance by regulated entities with federal pollution control requirements, with the imposition of strong enforcement actions in response to violations. Some critics have depicted environmental enforcement as overly litigious, or requiring unwarranted remedies. Others counter that actions are not pursued with enough rigor and frequency, or that penalties are not severe enough to deter noncompliance. EPA officials have countered that, in some instances, the agency is relying more on settlements and focusing on requiring increased expenditures on pollution control technologies, and that it is focusing judicial actions on larger and more complex cases that are expected to result in larger environmental benefits.

EPA and states maintain a considerable degree of flexibility in determining how to respond to potential violations, to the extent authorized by individual statutes. Initially, a potential violation is identified through monitoring, inspecting, citizen reporting, or through self-reporting by the regulated entity. As a first step in the enforcement process, unless an imminent danger or hazard has been determined, EPA and states may attempt to obtain corrective actions by simply issuing a warning or notifying a facility that minor violations may exist, and granting reasonable time for compliance. EPA or a state may then (or sometimes as a first step) initiate a civil administrative action under its own authority without involving the judicial process, or file formal civil or criminal[58] judicial actions in court.

Sanctions imposed, whether through negotiated settlements or decisions by the court, generally include required actions to achieve compliance and to correct environmental damage (injunctive relief), and may include monetary penalties (and incarceration in the case of criminal violations). More recently, settlements have also included requirements that violators undertake mutually agreed-upon environmentally beneficial projects supplemental to other sanctions.[59]

As noted, EPA, states, and the courts have considerable discretion in determining sanctions and remedies on a case-by-case basis so that the individual circumstances of each case are appropriately addressed. A majority of environmental violations are addressed and resolved administratively by states and EPA, and many of these cases are settled through negotiations between the government and the alleged violator. For example, during FY2008, EPA issued 1,390 administrative orders and filed 2,056 administrative penalty order complaints. In comparison, during FY2008, EPA filed 164 civil judicial cases with the court, and 192 civil judicial enforcement cases were concluded.[60] Civil judicial cases constitute the second largest category of environmental enforcement actions. Historically, judicial actions focused on violation of a single environmental statute. In recent years, EPA and states have increased the frequency of reliance on a multimedia (multi-statute) approach and multimedia investigations.

The number of administrative and judicial enforcement actions and penalties often fluctuate significantly from year to year. These fluctuations are generally a reflection of a combination of factors, including statutory deadlines; new or amended requirements in response to new scientific information or amended and new regulations; increased or decreased resources; environmental priority changes at the federal or state levels; and increased or improved monitoring/reporting. For example, EPA reported that the number of administrative penalty order complaints issued by the agency more than doubled, from 2,229 complaints in FY2005 to 4,647 in FY2006, then declined in FY2007 to 2,237 and to 2,056 in FY2008.[61] Additionally, the total dollar amount of penalties collected in a given year could reflect the completion of one or two large cases. For example, EPA reported that a single case accounted for 62% of the total civil penalties assessed for FY2004. Illustrations of the frequency of enforcement actions by type over time are presented in the **Appendix**; this appendix also includes illustrations of administrative and judicial penalties assessed over time by statute.

Monitoring, Inspections, and Evaluations

Critical steps in enforcing environmental laws include the compilation of monitoring data, and inspection and evaluation of the activities of the regulated community to determine who is complying with applicable regulatory requirements and permit conditions, and who is not. Compliance monitoring, evaluations, and investigations all serve to identify violations and provide insights into potential priority issue areas that may need to be addressed more broadly. Monitoring and reporting can be both media program-based (e.g., air, water, waste) and sector- based (e.g., industrial, mobile source, utilities), and are often included in permit requirements. Data reported and obtained, as well as observations and evidence collected by inspectors, enable EPA and states to identify specific environmental problems and determine whether a facility is in compliance. The information and evidence could eventually be used in an enforcement action. The mere collection of information or threat of inspection itself often creates an awareness of the regulators' interest, and can encourage compliance.

EPA identifies several forms of compliance monitoring that are used differently by the agency and states, depending upon the statute, the nature of the pollutants, and the types of facilities being regulated:

- **Self-Monitoring/Reporting:** Most environmental laws require (typically through permitting) regulated entities/facilities to monitor and record their own compliance status and report some or all of the tracking results to the responsible regulating authority. In addition to informing the regulators, self-monitoring also allows a company to measure its performance and evaluate its strategies for achieving or maintaining compliance.
-
- **Review of Records:** Regulatory agencies review data and information reported or otherwise compiled and collected.

Federal Pollution Control Laws 113

- **Full and Partial Inspections/Evaluations:**[62] Individual facility environmental inspections, conducted by EPA regional staff and the states, are the primary tool used by regulators for initial assessment of compliance. Through sampling, emissions testing, and other measures, inspections examine environmental conditions at a facility to determine compliance (or noncompliance) with specific environmental requirements, and to determine whether conditions present imminent and substantial endangerment to human health and the environment. Inspections/evaluations can be conducted all at once or in a series of partial inspections.

-

- **Area Monitoring:** Area monitoring looks at environmental conditions in the vicinity of a facility, or across a certain geographic area. Examples of methods used for area monitoring include ambient monitoring and remote sensing.

According to the EPA's most recent reported trends data, a total of 20,000 EPA enforcement inspections and evaluations were conducted under the various statutes during FY2008.[63] Although most inspections are carried out by the states, annual data for the total number of inspections conducted by states are not readily available due to data-reporting variability and other limitations. Based on a subset of states surveyed, ECOS reported that roughly 136,000 compliance inspections were conducted by states in 2003 for the major federal environmental programs—air, drinking water, surface and groundwater, hazardous waste, and solid waste.[64] The total number of inspections reported by ECOS does not account for all inspections conducted by states under federal pollution control programs—for example, inspections under FIFRA are not included. In reports to EPA by states under the Pesticide Enforcement Grant program, states, tribes and territories reported between 90,000 and 100,000 FIFRA inspections each fiscal year for FY2006 and FY2007. These FIFRA activities, typically administered by states' departments of agriculture, are not reflected in the EPA or the ECOS totals.

To put the ECOS number of inspections into perspective, in 2003, the ECOS survey identified 440,000 regulated facilities under these five major environmental programs. EPA's Facility Registry System (FRS), which identifies facilities and sites subject to federal environmental regulation, currently contains unique records for more than 1.5 million facilities (see the above discussion under the heading Regulated Community). The **Appendix** presents data on the number of inspections conducted annually by EPA over time.

Civil Administrative Actions

As noted earlier, a majority of environmental pollution control violations are addressed and resolved administratively by states and EPA without involving a judicial process. EPA or a state environmental regulatory agency may informally communicate to a regulated entity that there is an environmental problem, or it may initiate a formal administrative action in the form of a notice of violation or an Administrative Order to obtain compliance. An Administrative Order imposes legally enforceable requirements for achieving compliance, generally within a specified time frame, and may or may not include sanctions and penalties.

114   Robert Esworthy

An initial step in the enforcement process is often a Notice of Violation, or in some instances, a warning letter. Warning letters are issued mostly for first-time violations that do not present an imminent hazard. These notifications are intended to encourage regulated entities to correct existing problems themselves and come into compliance as quickly as possible. According to EPA, in many cases, these notices are not escalated to further formal enforcement action because a facility corrects problems and returns to compliance in response to the notice.

Through administrative enforcement actions, the EPA and states may (1) require that the violator take specific actions to comply with federal environmental standards, (2) revoke the violator's permit to discharge, and/or (3) assess a penalty for noncompliance. As indicated previously, administrative actions frequently end in negotiated settlements. These mutually agreed-upon resolutions are typically in the form of a Consent Agreement or Final Administrative Order/Penalty. According to EPA's FY2008 annual results, during FY2008, EPA initiated 1,390 administrative compliance orders and 2,056 administrative penalty order complaints. EPA imposed penalties in 2,084 final administrative penalty orders during FY2008, representing a total value of $38.2 million.[65]

Federal administrative orders are handled through an administrative adjudicatory process, filed before an administrative law judge (ALJ), or, in the regions, by EPA's regional judicial officers (RJOs). The EPA Office of Administrative Law Judges (OALJ) is an independent office within the agency.[66] ALJs, appointed by the EPA Administrator,[67] perform adjudicatory functions and render decisions in proceedings between the EPA and individuals, entities, federal and state government agencies, and others, with regard to administrative actions taken to enforce environmental laws and regulations. RJOs, designated by each of the EPA Regional Administrators,[68] perform similar adjudicatory functions in the EPA regions. Decisions issued by ALJs and RJOs are subject to review and appeal to the Environmental Appeals Board (EAB), which also functions independently of EPA.[69] Environmental Appeals Judges are appointed by the EPA Administrator.[70] Federal pollution control laws and regulations specify who may raise an issue before the EAB, and under what circumstances. EAB decisions frequently involve reviews of the terms of federal environmental permits and the amount of assessed financial administrative penalties.

Civil Judicial Enforcement

After civil administrative enforcement actions, civil judicial cases constitute the next-largest category of environmental enforcement. These are lawsuits filed in court against persons or entities who allegedly have not complied with statutory or regulatory requirements, or, in some cases, with an Administrative Order. Authorities for pursuing civil judicial actions and penalties are specified in each of the individual environmental statutes. Civil judicial cases are brought in federal district court by DOJ on behalf of EPA, and, for the states, by State Attorneys General. Not all of the cases referred to DOJ are filed with the court. The length of a civil case from its initiation to completion is highly variable, often extending across several years and sometimes across different presidential administrations. Like administrative enforcement actions, many civil judicial actions end as negotiated settlements, typically in the form of Consent Decrees. During FY2008, EPA-OECA referred

Federal Pollution Control Laws 115

280 civil judicial cases to DOJ; 164 civil judicial complaints were filed with the court; and 192 cases were concluded (refers to cases filed during FY2008 and prior fiscal years).

Criminal Judicial Enforcement

States and EPA may initiate criminal enforcement actions against individuals or entities for negligent or knowing violations of federal pollution control law. Criminal actions are especially pursued when a defendant knew, or should have known, that injury or harm would result. Knowing criminal violations of pollution control requirements are considered deliberate, and not the result of accident or error. In addition to the imposition of monetary fines and requirements to correct a violation and restore damages, conviction of a criminal environmental violation can result in imprisonment. EPA reported that 319 new environmental crime cases were opened during FY2008, about 5% less than in FY2007.[71] Authorities for pursuit of criminal actions vary under each of the statutes. For example, under the SDWA (42 U.S.C. §300h-2(b)), the criminal violations must be deemed willful—that is, they were committed with intent to do something prohibited by that law; the CWA (33 U.S.C. §1319(c)) authorizes criminal sanctions against those who have knowingly or negligently violated that statute.

Recent examples of criminal actions include the illegal disposal of hazardous waste; importation of certain banned, restricted or regulated chemicals; the export of hazardous waste without prior notification or permission of the receiving country; the removal and disposal of regulated asbestos-containing materials inconsistent with requirements of the law and regulations; tampering with a drinking water supply; and negligent maintenance resulting in discharge of hazardous materials.[72]

The EPA-OECA Office of Criminal Enforcement, Forensics, and Training (OCEFT), the office to which the agency's criminal investigators are primarily assigned, oversees implementation of the agency's federal environmental crimes investigation program. Within DOJ, the U.S. Attorneys Offices and ENRD's Environmental Crimes Section (ECS) prosecute criminal cases and work closely with EPA's OCEFT investigators. State and local law enforcement agencies and their environmental protection-related agencies, and other federal agencies, are also often key participants in federal environmental criminal actions. To facilitate investigations and cases, environmental crime task forces have been established nationally.[73] These task forces are composed of representatives from federal (including representatives from DOJ-ECS and special agents from EPA), state, and local law enforcement, and environmental regulatory enforcement. The FBI, DOT, Coast Guard, Fish and Wildlife Service, Army Corps of Engineers, SEC, IRS, and other relevant federal agencies also may play significant roles.

An increased emphasis on criminal enforcement of the pollution control laws occurred in the mid 1970s with the issuance of extensive guidelines for proceeding in criminal cases, and in 1981 with the creation of an Office of Criminal Enforcement and the hiring of criminal investigators in EPA's regional offices. During the late 1980s, criminal environmental enforcement was further enhanced when Congress conferred full law enforcement powers upon EPA criminal investigators as part of the Medical Waste Tracking Act of 1988 (18 U.S.C. §3063). Further, under Title II of the Pollution Prosecution Act of 1990 (P.L. 101-

593), Congress authorized the appointment of a director of a new Office of Criminal Investigations within EPA, and mandated the hiring of 200 criminal investigators by FY1 996.[74]

EPA's criminal enforcement agents are authorized law enforcement officers who, in addition to investigating federal environmental statutes, investigate U.S. Criminal Code (Title 18) violations often associated with environmental crimes, such as conspiracy, false statements, and interfering with federal investigations. As noted, Congress has been concerned with the staffing of criminal investigators. The number of EPA special agents went from about 50 in 1990 to more than 200 by 1998.[75] As of September 2007, the number of EPA investigators had dropped to 168; this decline was an issue of concern in Congress and elsewhere. EPA has indicated that as part of an OECA three-year hiring strategy, it plans to hire additional criminal investigators during FY2008, FY2009, and FY20 10, to increase the number of agents supporting the criminal enforcement program to 200.[76] **Table 3** shows the number of EPA investigators assigned to the criminal enforcement program for FY1997 through September FY2008, and projections for FY2009 and 2010, as reported by EPA.[77]

Table 3. Number of EPA Criminal Investigators: FY1997- FY2007

Fiscal Year	Number of Investigators
1997	200
1998	200
1999	192
2000	179
2001	181
2002	217
2003	217
2004	202
2005	189
2006	183
2007	168
2008	183
2009	192[a]
2010	200[a]

Source: Prepared by CRS with data provided by EPA's Office of Congressional and Intergovernmental Relations, Oct. 31, 2007, and March 26, 2008.

Note: The number of agents listed for 2002-2004 includes investigators who performed a mix of environmental crimes and homeland security work. The number of agents in FY2005 and FY2006 includes a cumulative total of 1.5 FTEs for specialized national security event response training of 25 agents so they can be deployed, if needed, to respond to a national security event.

a. Projected.

Sanctions[78] and Penalties

Settlements often require that violators achieve compliance and remedy environmental damages (injunctive relief). Monetary penalties may be included. Sanctions can also include permanent or temporary closure of facilities or specific operations, increased monitoring/reporting, revocation of existing permits or denial of future permits, and barring of receipt of federal contract funding or other federal assistance.[79] The settlement-required corrective and compliance actions, and the monetary penalties (and possibly incarceration for criminal violations), are intended to correspond directly with the specific violations (noncompliance) and the extent (or "gravity") of action committed.

Monetary penalties collected by the federal government as a result of an environmental enforcement agreement, order, or decision, are deposited with the U.S. Treasury.[80] However, under CERCLA (Superfund) and CWA, money recovered for the costs of replacing or restoring natural resources is used to restore the resources.[81] States may have the explicit administrative authority to impose penalties under individual federal statutes. For example, the Safe Drinking Water Act requires (unless prohibited by a state's constitution) administrative penalty authority for states in certain dollar amounts as a condition of obtaining and/or retaining primacy for the Public Water System Supervision (PWSS) Program (§1413 (a)(6)).[82] Currently, 55 of 57 states and territories have primacy authority for the PWSS program.[83] Although authorized under several of the other federal pollution control laws, EPA has not required—and not all states have obtained—administrative penalty authority. In some states, unlike the federal government, penalties obtained (or shared) as a result of an environmental enforcement action can be used to directly fund activities for environmental agencies and programs in the state, and not always to fund the state's general treasury.

In certain cases where the federal government has led the enforcement action, a state or states involved in the action may "share" resulting civil monetary penalties to the extent that the division is permitted by federal, state, and local law. A number of critical factors must be considered in accordance with EPA guidance[84] when determining division of penalties, including the state's active participation in prosecuting the case and its authority to collect civil penalties. EPA's guidance emphasizes that an agreement to include a division of civil penalties with states must be completed prior to issuance of a final settlement (order or consent decree).

The several statutes establish various factors to be considered in determining penalties: (1) the magnitude of environmental harm and the seriousness or gravity of a violation; (2) the economic benefit or gain to the violator as a result of illegal activity (noncompliance), including the gaining of a competitive advantage by the delaying or avoidance of pollution control expenditures that have been incurred by those in compliance; (3) violation history of the violator; and (4), in some circumstances, the ability of the violator to pay. Other factors, such as the degree of cooperation by the violator, whether the violation is self-reported, or the extent to which immediate action has been taken by the violator to mitigate potential harm, may also be considered. Precedents in previous cases involving similar violations are also a consideration when determining penalties.

The federal pollution control statutes include civil administrative and judicial penalty assessment authority and limits, which are to be considered by ALJs or the courts in determining the appropriate penalty. **Figure A-3** in the **Appendix** presents examples of dollar amounts of civil administrative, civil judicial, and criminal penalties assessed by EPA for the 20-year period FY1989-FY2008. According to EPA, a significant portion of the reported total annual dollar amount of all penalties assessed often reflects penalties assessed in a few cases, and, in some years, a single case. For example, EPA reported that a major RCRA case accounted for 26% of the total value of civil penalties reported in FY2006, and that, in FY2005, penalties assessed in a single RCRA corrective action case accounted for 53% of the reported assessed civil penalties for the year.[85]

EPA and DOJ have established several policies and guidelines to be considered by counsel when negotiating agreements and setting penalties.[86] EPA-OECA has also developed five computer models including models for calculating economic advantage, costs of Supplemental Environmental Projects (SEPs; see discussion under the heading "Supplemental Environmental Projects (SEPS)"), and for measuring the ability to afford compliance requirements and penalties. The latter models vary depending on whether a violator is an individual, municipality, individual facility, or business entity (small business, large corporation, or conglomerate partnership). Findings of limited ability or inability to pay are one factor under which an enforcement case may be settled for less than the economic benefit of noncompliance. The models are to be used in conjunction with the policies and guidelines for calculating civil penalties.[87] The available models are:

- BEN, for calculating economic advantage/savings from avoidance of compliance;
- ABEL, for measuring a noncompliant entity's (e.g., a corporation's) ability to afford compliance and cleanup, and civil penalties;
- INDIPAY, for measuring an individual violator's ability to afford compliance and cleanup, and civil penalties;
- MUNIPAY, for measuring a noncompliant municipality's ability to afford compliance and cleanup, and civil penalties; and
- PROJECT, for calculating cost to a violator of undertaking a SEP (see the discussion regarding SEPs later this chapter).

Penalties Assessed to Federal Facilities[88]

Most federal pollution control statutes contain a provision expressly subjecting federal facilities to federal (and state and local) environmental regulation, and waiving sovereign immunity (thereby allowing federal agencies to be sued by nonfederal entities). Further, many federal environmental statutes authorize (or arguably authorize) EPA, states, and local governments to assess civil monetary penalties against federal agencies. (The Supreme Court has rejected state authority to do so under the CWA.[89]) DOJ has issued opinions concluding that the CAA and RCRA underground storage tank provisions give EPA authority to assess civil money penalties against federal facilities. However, DOJ limits these conclusions to *administrative* assessment of penalties. Citing its constitutional theory of the "unitary executive," DOJ has historically refused to allow EPA to enforce *judicially* against other federal agencies, though case law has consistently been to the contrary. In contrast with EPA enforcement, there is no longer serious doubt that the Constitution allows states and other nonfederal entities to use the citizen suit provisions in federal environmental statutes to judicially enforce those laws against federal facilities.

During FY2008, EPA concluded 36 enforcement actions against federal facilities, and assessed more than $1.4 million in penalties. Federal agencies committed to invest more than $23 million to improve their facilities and operations to remedy (clean up) past violations.[90]

Supplemental Environmental Projects (SEPs)

In addition to requiring violators to achieve and maintain compliance, and imposing appropriate sanctions and penalties, enforcement settlements may also include Supplement Environmental Projects (SEPs).[91] SEPs are projects that provide environmental and human health benefits that a violator may voluntarily agree to undertake in exchange for mitigation of penalties. A project must be related to the violation, and cannot be an activity the violator is legally required to take to achieve compliance. Penalties are to be mitigated by a SEP only during settlement negotiation, prior to imposition of the final penalty.

EPA has established a SEPs policy and developed guidance for their legal requirements and applicability, and has specified eight categories of acceptable projects.[92] These include pollution prevention, public health, and emergency and preparedness planning. EPA reported that 188 civil settlement cases during FY2008 included SEPs at an estimated value of $39.0 million.[93]

The incorporation of SEPs into enforcement actions has become a more commonly used enforcement tool in recent years, particularly by federal regulators, because of the potential for direct environmental benefit from such projects, versus the use of a monetary fine or penalty alone. Some states with administrative penalty authority are also employing the use of SEPs in their settlements.[94] Although these projects are required to be supplemental to other requirements, some contend that, in practice, inclusion of SEPs may result in lower monetary fines. The extent to which specific SEPs may have resulted in reduced monetary fines and penalties is not easily calculable.

Environmental Justice and Enforcement/Compliance

The concept of "environmental justice" has been a controversial topic of debate among industry and public interest groups, and continues to be a concern highlighted in legislation and congressional hearings. During the 110th Congress, hearings have been held by committees in the House and the Senate,[95] and legislation has been introduced, including H.R. 1972, the Community Environmental Equity Act; H.R. 4652, the Environmental Justice Access and Implementation Act of 2007; companion bills H.R. 1103 and S. 642, the Environmental Justice Act of 2007, H.R. 5132 and S. 2549, the Environmental Justice Renewal Act, and, H.R. 5896 and S. 2918, the Environmental Justice Enforcement Act. Discussion of the full scope of issues and concerns regarding environmental justice is beyond the scope of this chapter. However, the following discussion briefly highlights environmental justice in the context of enforcement and compliance.

The terms "environmental justice (or injustice)" and "environmental equity (or inequity)" may be interpreted broadly to describe the perceived level of fairness in the distribution of environmental quality across groups of people with different characteristics. In this sense, the environmental impact of any human activity might be evaluated to determine the distribution of environmental amenities and risks among people categorized according to any population characteristic, including gender, age, race, place of residence, occupation, income class, or language. In the political context, however, emphasis generally is more on the distribution of health risks resulting from exposure to toxic substances in residential or occupational environments of different racial, ethnic, or socioeconomic groups.

The 1994 Executive Order 12898, Federal Actions to Address Environmental Justice in Minority Populations and Low-Income Populations, directs each federal agency to "make achieving environmental justice part of its mission."[96] EPA is the federal agency with lead responsibility for implementing the executive order. EPA's Office of Environmental Justice (OEJ), located in OECA, is responsible for coordinating efforts to include environmental justice into policies and programs across the agency's headquarters and regional offices.[97] EPA's OEJ provides information and technical assistance to other federal agencies for integrating environmental justice into their missions, engages stakeholders to identify issues and opportunities, and administers EPA environmental justice grants.[98]

EPA's OEJ is developing the Environmental Justice Strategic Enforcement Assessment Tool (EJSEAT), which it expects to begin using in FY2008. OECA expects to use EJSEAT to "consistently identify possible environmental justice areas of concern," where potentially disproportionately high and adverse environmental and public health burdens exist, and assist EPA in making "fair" enforcement and compliance resource deployment decisions.[99] OEJ published a "Toolkit for Assessing Potential Allegations of Environmental Injustice," primarily to assist agency staff in assessing allegations of environmental injustice.[100] Citizens can evaluate overlap between environmental conditions and demographic characteristics by using EPA's Environmental Justice Geographic Assessment Tool,[101] through the agency's online database— *EnviroMapper* (see the discussion regarding EPA's various enforcement compliance databases in the **Appendix**).

Compliance Assistance and Incentive Approaches

A frequent criticism regarding implementation and enforcement of federal environmental requirements has been an emphasis, historically, on a "command and control" approach. In response to these criticisms, since the 1990s EPA and states have relied increasingly on compliance assistance to help the regulated community understand its obligations to prevent violations and reduce the need for enforcement actions, as well as to assist violators in achieving compliance. Many states have advocated compliance assistance and developed assistance programs designed to address specific environmental issues at the local level.[102]

EPA's Office of Compliance (OC) within OECA has introduced a number of compliance assistance programs, many of them developed in conjunction with support from the regions, states, and tribes.[103] Each EPA region has a designated Compliance Assistance Coordinator who serves as an "expert" within the region on compliance assistance priorities, strategies, and performance measurement. The coordinators work with subject-matter experts in the regions and at headquarters in the development of compliance assistance guides and workshops, and contribute to other assistance activities such as conducting compliance assistance visits.[104]

In addition to providing compliance assistance across the individual pollution control statutes, sector-based assistance is also provided. Developed and introduced in partnership between EPA, states, academia, environmental groups, industry, and other agencies, the National Compliance Assistance Centers provide sector-specific assistance.[105] There are currently 15 sector-specific Web-based compliance assistance centers. As shown in **Table 4** below, the sector-specific centers include agriculture, auto repair, chemical manufacturing, federal facilities, and local governments.

Through partnerships, EPA has developed the National Compliance Assistance Clearinghouse,[106] providing Web-based access to compliance tools and contacts, in order to facilitate information- sharing on compliance assistance. Trade associations, universities, and consultants are also an increasing source of assistance information.

The use of compliance incentive approaches has been evolving. Incentives generally are policies and programs that may reduce or waive penalties and sanctions under specific conditions for those who voluntarily take steps to evaluate, disclose, correct, and prevent noncompliance. Examples include self-disclosure programs and related tools such as environmental audit protocols, Environmental Management Systems, and other innovation projects and programs designed to achieve environmental benefits.

One of the earliest formal EPA incentive approaches is the EPA Audit Policy— "Incentives for Self-Policing: Discovery, Disclosure, Correction and Prevention of Violations"—in effect since 1995.[107] Under the policy, certain violations are voluntarily reported after being discovered through self-audit. In many cases EPA eliminates civil penalties, and may offer not to refer certain violations for criminal prosecution. In early 2007, EPA solicited comments on the question of to what extent, if any, the agency should consider providing incentives to encourage new owners of recently acquired facilities to discover and disclose environmental violations, and to correct or prevent their reoccurrence.[108] More recently, to further promote compliance through the use of various incentive approaches, EPA has encouraged incentive approaches as part of its core program guidance included in the OECA FY2009 National Program Manager Guidance. The guidance was distributed to Regional Administrators and State Environmental Commissioners in June 2008.[109]

Table 4. Sector Web-Based Compliance Assistance Centers

Sectors	
Agriculture	Local Government
Automotive Recycling	Metal Finishing
Automotive Service and Repair	Paints and Coatings
Chemical Manufacturing	Printed Wiring Board Manufacturers
Colleges/Universities	Printing
Construction	Transportation
Federal Facilities	Tribal Governments and Indian Country
Healthcare	

Source: Table created by CRS with information from the National Compliance Center website, available at http://www.assistancecenters.net/.

EPA's reliance on incentive approaches has been met with some skepticism by those who favor more traditional enforcement. Critics are concerned that incentive and voluntary approaches subtract resources from an already limited pool of enforcement resources. EPA and other supporters of these approaches contend that they result in cost savings by reducing burdens on investigators, achieve desired environmental improvements, and allow for the leveraging of additional resources through partnerships. Aspects of EPA's incentive approaches have been the subject of reviews by EPA-OIG and GAO.[110]

FUNDING FOR ENFORCEMENT/COMPLIANCE ACTIVITIES

The adequacy of resources needed by EPA, DOJ, and the states to effectively enforce the major federal environmental pollution control laws is often highlighted during congressional debate of fiscal year appropriations. Congress has specified funding levels for certain aspects of EPA enforcement activities, or required the agency to undertake certain actions under annual appropriations; an example is the previously mentioned provision included in the House-passed FY2008 Interior and Environmental Agencies Appropriations bill (H.R. 2643) that would have required EPA to hire criminal investigators to bring the total number of investigators up to the statutory requirement of 200, pursuant to the Pollution Prosecution Act of 1990.[111] (The provision was not included in the FY2008 consolidated appropriations.[112])

The FY2008 enacted appropriation for EPA's enforcement activities was $553.5 million.[113] **Table 5** illustrates the distribution of funding and full-time equivalents (FTEs) among various enforcement activities across the agency's appropriations accounts for the two most recently completed fiscal years. (Similarly detailed information on the distribution of funding for other fiscal years, including FY2008, is not readily available.)

Table 5. EPA-OECA's FY2007-FY2008 Enacted Appropriation and FTEs by EPA Appropriations Account and Program Activity (dollars in thousands)

EPA Appropriation Account / Program Activity	FY2007 Enacted		FY2008 Enacted	
	Dollars	FTEs	Dollars	FTEs
Total	**$548,934.0**	**3,450.0**	**$553,543.0**	**3,415.2**
Environmental Programs Management (EPM)	**$322,933.0**	**2,241.2**	**$326,943.0**	**2,242.7**
Brownfields	$768.0	5.0	$778.0	4.6
Civil Enforcement	$125,438.0	942.8	$129,886.0	964.9
Compliance Assistance and Centers	$29,070.0	204.8	$27,725.0	196.8
Compliance Incentives	$9,695.0	75.7	$10,618.0	73.7
Compliance Monitoring	$92,769.0	630.1	$88,726.0	619.6
Congressionally Mandated Projects	$0.0	0.0	$0.0	0.0
Criminal Enforcement	$38,939.0	221.6	$40,742.0	220.1
Enforcement Training	$2,562.0	12.8	$3,096.0	15.6
Environmental Justice	$4,675.0	16.9	$6,399.0	16.9
Homeland Security	$4,006.0	20.8	$3,904.0	20.8
International Capacity Building	$0.0	0.0		
NEPA Implementation	$13,967.0	104.0	$14,142.0	104.0
Congressional, Intergov., Ext. Rel.	$1,044.0	6.7	$927.0	5.7

Federal Pollution Control Laws

Table 5. (Continued)

EPA Appropriation Account / Program Activity	FY2007 Enacted		FY2008	
	Dollars	FTEs	Dollars	FTEs
IT / Data Management	$0.0	0.0		
Science and Technology (S&T)	**$13,568.0**	**83.0**	**$14,882.0**	**90.5**
Forensics Support	$13,568.0	83.0	$14,882.0	90.5
State and Tribal Assistance Grants (STAG)	**$25,913.0**	**—**	**$24,647.0**	**—**
Categorical Grant: Pesticides Enforcement	$18,622.0	—	$18,419.0	—
Categorical Grant: Toxics Substances	$5,074.0	—	$5,019.0	—
Categorical Grant: Sector Program	$2,217.0	—	$1,209.0	—
Leaking Underground Storage Tanks (LUST)	**$724.0**	**5.5**	**$709.0**	**4.8**
Compliance Assistance and Centers	$724.0	5.5	$709.0	4.8
Oil Spills Response	**$2,007.0**	**15.8**	**$2,358.0**	**17.3**
Civil Enforcement	$1,730.0	14.0	$2,072.0	15.5
Compliance Assistance and Centers	$277.0	1.8	$286.0	1.8
Hazardous Substances Superfund	**$183,789.0**	**1,104.5**	**$184,004.0**	**1,059.9**
Civil Enforcement	$880.0	1.7	$870.0	1.7
Compliance Assistance and Centers	$22.0	0.0	$22.0	0.0
Compliance Incentives	$141.0	0.9	$159.0	0.9
Compliance Monitoring	$1,182.0	1.9	$1,165.0	1.9
Criminal Enforcement	$9,047.0	49.2	$9,053.0	48.8
Enforcement Training	$612.0	4.1	$827.0	5.3
Environmental Justice	$757.0	0.0	$745.0	0.0
Forensics Support	$3,802.0	24.8	$3,750.0	15.3
Homeland Security	$1,813.0	9.2	$1,828.0	9.2
Superfund: Enforcement	$155,021.0	930.3	$155,705.0	901. 4
Congressional, Intergov., Ext. Rel.	$151.0	1.1	$154.0	1.1
Superfund Federal Facilities Enforce.	61.0	81.3	6.0	74.3

Source: Compiled by the Congressional Research Service with data received from the Environmental Protection Agency's Office of Congressional and Intergovernmental Relations (OCIR) in written communications: June 4, 2007 and February 12, 2008.

DOJ's resource (funding/staff) requirements and outlays associated with its litigation activities under the major federal pollution control statutes are, in the main, a subset of the funding (proposed and previously appropriated) for ENRD in its annual budget justifications. As discussed previously, ENRD is responsible for the majority of DOJ's support of the federal pollution control laws, as well as many other responsibilities, including representing the United States in matters regarding natural resources and public lands, acquisition of real property by eminent domain for the federal government, and cases under wildlife protection laws.

The President's FY2009 budget request for DOJ included $103.1 million and 499 FTEs for ENRD. An additional $23.9 million for 184 FTEs included in the President's request for EPA was to be transferred to ENRD through a reimbursable agreement for Superfund work. FY2008 enacted levels for ENRD were $99.5 million and 495 FTEs, plus $26.2 million for 184 FTEs transferred from EPA. Of the FY2008 enacted amount including the transfers from EPA, roughly $67.7 million and 366 FTEs were for environmental litigation activities; $10.9 million and 59 FTEs were for criminal litigation conducted by ECS; and $56.8 million and 307 FTEs were for civil environmental defensive and enforcement litigation conducted by the Environmental Defense and the Environmental Enforcement Sections.[114]

Detailed reporting of federal funding to states and states' funding contributions for pollution control enforcement/compliance activities is not readily available. ECOS has tracked a broader category of state funding and expenditures that it defines as annual "environmental and natural resource spending," which, in more recent years, has been primarily based on survey data reported by states. The data, which include state and federal funding, are limited for purposes of enforcement of federal pollution control laws in that they combine environmental and natural resource spending. Also, states vary in how they track and report this type of spending. The data do provide a source of state funding from a national perspective. For example for FY2003, the most recent fiscal year reported, ECOS reported that states budgeted a combined total of $15.0 billion for environment and natural resources spending; this represents 1.4 percent of their combined total budgets. ECOS found that $5.0 billion, or one-third of the amount budgeted, was from federal funding to states for these purposes.[115]

Federal appropriations, in particular allocations to states, for adequate staffing and effective enforcement of federal environmental statutes to protect human health and the environment, will likely continue to be an issue of concern.

CONCLUSION

Fully evaluating and measuring the overall effectiveness of current (and past) enforcement/compliance activities can be quite complicated. Discussion throughout this chapter highlights the difficulties inherent in characterizing the many facets of environmental enforcement at a macro level, and identifies many of the factors that may contribute to its perceived successes and shortcomings. However, several indicators do provide insight into a better understanding of the complexities associated with elements of enforcement, such as the vastness and diversity of the regulated community, the multiplicity of the activities and priorities across many regulating entities, and variability across statutes.

Since the establishment of EPA in 1970, Congress has been interested in a number of crosscutting issues associated with the enforcement of pollution control statutes and regulations, as reflected in provisions of enacted and amended environmental legislation over time. Congressional interest remains heightened, particularly with regard to the substance of intergovernmental relations, EPA-state relations, and fiscal requirements. Congress's involvement with these issues could take several directions. One likely result could be oversight hearings. Alternatively, relevant appropriations legislation may contain provisions or language regarding funding for specific enforcement activities.

Congressional interest might focus on statutory approaches to establish changes in the EPA-states' partnership, such as legislation similar to past proposals concerning refinement of the National Environmental Performance Partnership System (performance partnerships, or NEPPS) and the associated grants award process. Congress may also consider other statute-specific legislation to address other longstanding concerns that affect enforcement/compliance activities.

The regulated community, public interest groups, federal and state officials, and Congress are often divided on whether to pursue legislation that would further expand or constrain enforcement/compliance. They are similarly divided with respect to proposals that would

Federal Pollution Control Laws 125

expand states' authority for implementing and enforcing certain aspects of the major federal pollution control laws.

Views and congressional involvement with respect to these issues are likely to evolve in the years ahead.

APPENDIX. ENFORCEMENT/COMPLIANCE DATABASES AND EXAMPLES OF REPORTED RESULTS

Enforcement/Compliance Databases and Reporting

Compliance monitoring data are used to manage the compliance and enforcement program, and to inform the public of enforcement actions taken and penalties imposed. EPA and the states collect and maintain compliance/enforcement data in many forms. ECOS reports that states collect about 94% of environmental quality data contained in EPA's databases, primarily from state-issued permits and monitoring programs.[116] Information is often entered into multiple databases or transferred from state databases. Historically, the databases were often incompatible, making cross media/statute queries difficult. In recent years, EPA has been working to integrate several of the individual databases to allow more cross referencing of compliance data by regulators and to provide querying capabilities to the public.[117]

EPA compiles data from the various databases and provides various statistics in the form of annual accomplishment and multi-year trends reports.[118] Reporting has traditionally focused on statute-by-statute results, including actions initiated and concluded, and penalties and other sanctions assessed. How effectively the reported information can be used as an indicator of environmental progress and the impacts of environmental enforcement has been an issue of some debate, and questioned in reviews conducted by EPA-OIG and GAO.[119] Critics contend, and EPA has long recognized, that while somewhat indicative of the failure to comply with environmental requirements, counting enforcement actions alone ("bean counting") does not provide a complete measure of the effectiveness of the national environmental enforcement/compliance program.

EPA has initiated efforts to expand its reporting by including estimates of environmental benefits (pollution reduction and impacts avoided). Additionally, there have been efforts to account for states' contributions; the FY2006 OECA Accomplishment Report for the first time contained a brief summary of states' enforcement accomplishments as reported by ECOS.

Overview of Enforcement/Compliance Databases

A number of EPA's single- and multi-media national databases include enforcement and compliance data elements. While these databases are generally available to EPA staff, and in some cases state and local governments, most are not readily available to the public. The Enforcement and Compliance History Online, or ECHO, developed and maintained by OECA is the most prominent publicly accessible database. Introduced in 2003, ECHO queries

126 Robert Esworthy

provide a snapshot of the most recent three years of a facility's environmental compliance record, but are limited primarily to certain requirements under the CAA, CWA, and RCRA. EPA continues to expand the integration and capabilities of this and other databases. Finally, several state environmental agencies maintain additional information about compliance and enforcement (beyond what is reported to EPA systems).[120]

The following brief summaries of several of EPA's integrated national databases are a consolidation of descriptions provided on the agency's website:

Enforcement and Compliance History Online (ECHO)

ECHO is an interactive website that allows users to query permit, inspection, violation, enforcement action, informal enforcement action, and penalty information for individual or multiple facilities. Initial queries return a list of relevant facilities, each linked to a "Detailed Facility Report," indicating:

- whether a facility has been inspected/evaluated,
- occurrence and nature of violations (noncompliance),
- nature of enforcement actions (including penalties) that have been taken,
- contextual information about the demographics surrounding the facility.

Envirofacts

Envirofacts provides public access to information about environmental activities, such as releases, permit compliance, hazardous waste handling processes, and the status of Superfund sites, that may affect air, water, and land anywhere in the United States. Data are retrieved from various EPA source databases. Users can develop on-line queries, create reports and map results. See http://www.epa.gov/enviro/.

Facility Registry System (FRS)

FRS (a companion to the integrated facility searches in Envirofacts) can be used to create facility identification records, including geographical location, and to locate sites or places subject to environmental regulations or oversight (e.g., monitoring sites). Records are based on information from EPA program national systems, state master facility records, and data collected from EPA's Central Data Exchange. See http://www.epa.gov/frs/.

Integrated Compliance Information System (ICIS)

ICIS integrates data that are currently located in several separate data systems. ICIS contains information on federal administrative and federal judicial cases under the following environmental statutes: the CAA, CWA, RCRA, EPCRA, TSCA, FIFRA, CERCLA (Superfund), SDWA, and MPRSA. ICIS also contains information on compliance assistance activities conducted in EPA regions and headquarters. The Web-based system enables states and EPA to access integrated enforcement and compliance data. The public can only access some of the federal enforcement and compliance information in ICIS by using the EPA Enforcement Cases Search and EPA Enforcement SEP Search through ECHO. See http://www.epa.gov/compliance/ data/systems/modernization/index.html.

Integrated Data for Enforcement Analysis (IDEA)

IDEA maintains copies of EPA's air, water, hazardous waste and enforcement source data systems that are updated monthly. An internal EPA database, IDEA uses "logical" data integration to provide a historical profile of inspections, enforcement actions, penalties assessed and toxic chemicals released, for EPA-regulated facilities. See http://www.epa.gov/ compliance systems/multimedia/idea/index.html.

Online Tracking Information System (OTIS)

OTIS is a collection of search engines which enables EPA, state/local/tribal governments and certain other federal agencies to access a broad range of data relating to enforcement and compliance. No public access is available. This Web application sends queries to the IDEA system (discussed above). IDEA copies many EPA and non-EPA databases, and organizes the information to facilitate cross-database analysis. See http://www.epa.gov/compliance/data/ systems/multimedia/aboutotis.html.

A number of other databases, mostly for single media, also include compliance/ enforcement data. Many of these databases are the basis for certain data elements in the various integrated databases, and typically are not directly available to the public. Other databases include

- the Air Facility System (AFS);
- Permit Compliance System (PCS);
- Resource Conservation and Recovery Act Information System (RCRAInfo);
- National Compliance Data Base System and Federal Insecticide, Fungicide, and Rodenticide Act/Toxic Substances Control Act Tracking System (NCDB/FTTS); and
- Safe Drinking Water Information System/Federal (SDWIS/FED).

For a more complete list and descriptions of EPA's enforcement/compliance databases, see EPA's "Compliance and Enforcement Data Systems" Web page at http://www.epa.gov/ compliance/data/ systems/index.html.

Examples of Reported Enforcement Actions and Penalties Over Time

The following figures and tables provide examples of the type of enforcement data collected, compiled and reported over time. They are intended to show proportional relationships of the various types of enforcement actions (e.g., administrative vs. judicial) in a given year and by statute, not annual or long-term enforcement trends. To compare the reported activities from year to year requires more detailed information regarding the specific circumstances in those years. There can be significant variability from year to year in how data were reported and which entities reported. EPA has refined terms and definitions in the data elements from year to year. Other factors that result in variability include the introduction of new regulatory requirements in a given year, and fruition of statutory deadlines.

The figures presented below reflect longer-term data (15 to 20 years), whereas the tables generally provide data for the most recent five or six years, depending on the availability of data for the most recent fiscal year. FY2008 results data are available by action (e.g., administrative, civil judicial, criminal judicial), but not for actions by statute (e.g., Clean Water Act, Clean Air Act).

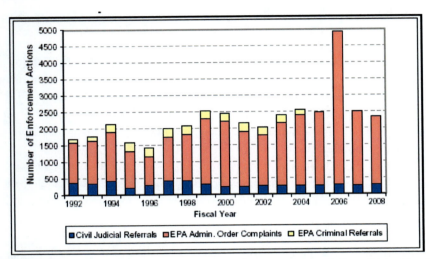

Source: Compiled by CRS using data from EPA's Enforcement Annual Compliance Results FY2008, National Enforcement Trends (FY2007)—EPA Long Term Trends; and National Enforcement Trends (FY2004)—EPA Long term Trends: Criminal Referrals and Penalties, http://cfpub.epa.gov/complianceindex.cfm. EPA terminated the count of criminal referrals as an internal Criminal Enforcement program measure in FY2005 and discontinued reporting this data in its trends reports

Figure A-1. EPA Civil Judicial Referrals, Administrative Order Complaints, and Criminal Referrals, FY1992-FY2008

Table A-1. EPA Civil Administrative, Civil Judicial, and Criminal Enforcement Actions, FY2003-FY2008

Enforcement Action	FY2003	FY2004	FY2005	FY2006	FY2007	FY2008
Administrative Compliance Orders	1,582	1,807	1,916	1,438	1,247	1,390
Administrative Penalty Order Complaints	1,888	2,122	2,229	4,647	2,237	2,056
Final Administrative Penalty Orders	1,707	2,248	2,273	4,624	2,256	2,084
Civil Judicial Referral	268	268	259	286	278	280
Civil Judicial Cases Concluded	195	176	157	173	180	192
Criminal Judicial Referral	228	168	NR	NR	NR	NR
Criminal Judicial Cases Initiated	508	422	372	305	340	319

Source: Compiled by CRS using data from EPA's Enforcement Annual Compliance Results FY2008, National Enforcement Trends (FY2007)—Enforcement Actions, and National Enforcement Trends (FY2004)—Enforcement Actions: Criminal Enforcement Program Activities, http://cfpub.epa.gov/compliance reports/nets/index.cfm.

Notes: NR = not reported. EPA terminated the count of criminal referrals as an internal Criminal Enforcement program measure in FY2005 and discontinued reporting this data in its trends reports.

Federal Pollution Control Laws

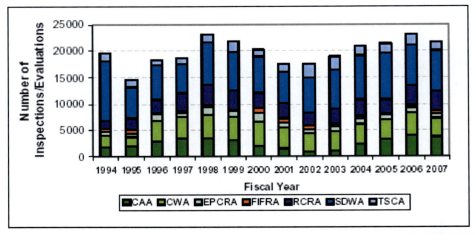

Source: Compiled by CRS using data from EPA's *National Enforcement Trends (FY2007)—EPA Long Term Trends: FY1994-FY2007 Federal Inspections and Evaluations*, http://cfpub.epa.gov/compliance/resources/reports/nets/ index.cfm.

Figure A-2. Number of EPA Federal Inspections and Evaluations by Statute, FY1994-FY2007

Table A-2. Number of EPA Enforcement Inspections and Evaluations by Statute, FY2002-FY2007

Statute	FY2002	FY2003	FY2004	FY2005	FY2006	FY2007
CAA	809	1,019	2,524	3,362	3,918	3,843
CWA	3,553	3,769	3,726	3,650	4,453	3,480
EPCRA	733	987	970	905	928	884
FIFRA	810	338	472	215	344	360
MPRSA	NR	3	4	0	0	0
RCRA	2,480	2,970	3,042	2,768	3,812	3,874
SDWA	6,545	7,418	8,501	8,771	7,768	7,618
TSCA	2,738	2,376	1,792	1,611	2,008	1,662
Total	17,668	18,880	21,031	21,282	23,231	21,721

Source: Compiled by CRS using data from EPA's National Enforcement Trends (FY2007)—EPA Inspections and Investigations, http://cfpub.epa.gov/compliance reports/ nets/ index. cfm. FY2008 were not available.

CAA: Clean Air Act
CWA: Clean Water Act
EPCRA: Emergency Planning and Community Right-to-Know Act
FIFRA: Federal Insecticide, Fungicide and Rodenticide Act
MPRSA: Marine Protection, Research, and Sanctuaries Act
RCRA: Resource Conservation and Recovery Act
SDWA: Safe Drinking Water Act
TSCA: Toxic Substances Control Act

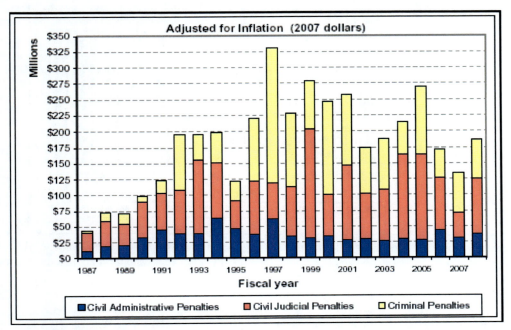

Source: Compiled by CRS using data from EPA's Enforcement Annual Compliance Results FY2008, *National Enforcement Trends (FY2007)—EPA Long Term Trends: FY1974-FY2007 Enforcement Penalties,* http://cfpub.epa.gov/compliance/resources/reports/nets/index.cfm. Amounts converted to 2007 dollars using the GDP Chained Price Index from the Office of Management and Budget, *Budget of the U.S. Government for FY2009,* Historical Tables, http://www.whitehouse.gov/omb/budget/fy2009/.

Figure A-3. Environmental Enforcement Penalties Assessed by EPA: Administrative, Civil Judicial, and Criminal, FY1986-FY2008.

Table A-3. Environmental Enforcement Penalties Assessed by EPA: Administrative, Civil Judicial, and Criminal, FY2003-FY2008

(dollars in thousands—not adjusted for inflation)

Fiscal Year	Administrative	Civil Judicial	Criminal[a]	Total
FY2003	$24,376	$72,260	$71,000	$167,636
FY2004	$27,637	$121,213	$47,000	$195,850
FY2005	$26,731	$127,206	$100,000	$253,937
FY2006	$42,007	$81,808	$43,000	$166,815
FY2007	$30,700	$39,800	$63,000	$133,500
FY2008	$38,200	$88,400	$63,500	$190,100

Source: Compiled by CRS using data from EPA's Enforcement Annual Compliance Results FY2008, *National Enforcement Trends (FY2007)—Penalties, Injunctive Relief and SEPs,* http://cfpub.epa.gov/compliance reports/nets/index.cfm.

a. Criminal penalties represent fines and restitution.

Federal Pollution Control Laws

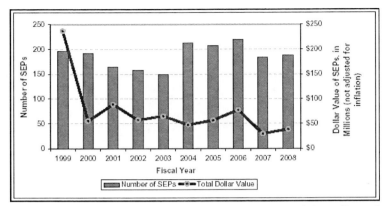

Source: Compiled by CRS using data from EPA's Enforcement Annual Compliance Results FY2008, National Enforcement Trends (FY2007)—Penalties, Injunctive Relief and SEPs: FY1999-FY2007 Supplemental Environmental Projects, http://cfpub.epa.gov/compliance/resources/reports/nets/index.cfm.

Figure A-4. EPA Supplemental Environmental Projects: Number of Projects and Dollar Value, FY1999-FY2008

Table A-4. Supplemental Environmental Projects (SEPs) Dollar Values as Reported by EPA: FY2002-FY2007

(dollars in thousands—not adjusted for inflation)

Statute	FY2002	FY2003	FY2004	FY2005	FY2006	FY2007
CAA	$33,109.9	$36,322.9	$21,816.0	$31,637.7	$41,712.9	$16,458.4
CERCLA	$2,960.2	$207.0	$564.8	$363.1	$2,732.8	$367.0
CWA	$13,078.7	$7,979.9	$16,067.0	$18,427.9	$21,712.8	$8,431.4
EPCRA	$1,223.3	$4,638.9	$1,094.4	$1,426.0	$1,208.2	$1,387.4
FIFRA	$12.0	$0.0	$247.4	$62.1	$31.1	$369.0
MPRSA	NR	$0.0	$0.0	$104.0	$0.0	$146.5
RCRA	$6,261.5	$15,218.1	$2,757.9	$3,249.5	$2,923.3	$1,720.8
SDWA	$428.2	$0.0	$322.9	$734.6	$133.1	$59.2
TSCA	$832.7	$1,054.7	$5,031.5	$1,031.1	$7,313.1	$1,405.0
Total	$57,906.3	$65,421.6	$47,901.9	$57,036.1	$77,767.3	$30,344.8

Source: Compiled by CRS using data from EPA's National Enforcement Trends (FY2007)—Penalties, Injunctive Relief and SEPs: FY1999-FY2007 Supplemental Environmental Projects, http://cfpub.epa.gov/ compliance resources/reports/nets/index.cfm. FY2008 data by statute was not available.

CAA: Clean Air Act
CERCLA: Comprehensive Environmental Response, Compensation and Liability Act (Superfund)
CWA: Clean Water Act
EPCRA: Emergency Planning and Community Right-to-Know Act
FIFRA: Federal Insecticide, Fungicide and Rodenticide Act
MPRSA: Marine Protection, Research, and Sanctuaries Act
RCRA: Resource Conservation and Recovery Act
SDWA: Safe Drinking Water Act
TSCA: Toxic Substances Control Act

End Notes

[1] See CRS Report RL3 0798, *Environmental Laws: Summaries of Major Statutes Administered by the Environmental Protection Agency (EPA)*.

[2] See, for example: Senate Committee on Environment and Public Works, "*Oversight Hearing on EPA Regional Inconsistencies,*" June 28, 2006, http://epw.senate Government Accountability Office (GAO): *Environmental Enforcement: EPA Needs to Improve the Accuracy and Transparency of Measures Used to Report on Program Effectiveness*, GAO-08-1111R, September 18, 2008; *Environmental Compliance and Enforcement: EPA's Effort to Improve and Make More Consistent Its Compliance and Enforcement Activities*, GAO- 06-840T, June 28, 2006; *Environmental Protection: More Consistency Needed Among EPA Regions in Approach to Enforcement*, June 2000, GAO/RCED-00-108, all available at http://www.gao.gov; EPA's Office of Inspector General (EPA-OIG): *Enforcement—Compliance with Enforcement Instruments*, Rpt. 2001-P-00006, March 29, 2001, http://www.epa.gov/oig/.

[3] Ibid.; see also GAO: Environmental Protection: Collaborative EPA-State Effort Needed to Improve Performance Partnership System, GAO/T-RCED-00-163, May 2, 2000, and. Environmental Protection: Overcoming Obstacles to Innovative State Regulatory Programs, GAO-02-268, January 31, 2002. See also EPA-OIG (http://www.epa.gov/oig/): EPA Needs to More Actively Promote State Self Assessment of Environmental Programs, Report No. 2003-P-00004, December 27, 2002.

[4] Many references discuss "cooperative federalism" in the context of environmental policy; these include Robert L. Fischman, *Cooperative Federalism and Natural Resources Law*, New York University Envtl. L. J. 179, vol. XIV 2006, Issue 1; Mark Agrast, et al., *How to Protect Environmental Protections?*, Envtl. Law Reporter, vol. 35, 2005 (10413 - 10417), the Environmental Law Institute; Philip J. Weiser, *Towards a Constitutional Architecture for Cooperative Federalism*, North Carolina L. Rev., vol. 79, 2001 (663, 671), University of North Carolina; Vickie L. Patton, *A Balanced Partnership*, The Envtl. Law Forum, vol. 13, no. 3, May/June 1996; and, Robert V. Percival, *Environmental Federalism: Historical Roots and Contemporary Models*, Maryland Law Rev., vol. 54, 1995 (1141).

[5] GAO, *Environmental Protection: EPA-State Enforcement Partnership Has Improved, but EPA's Oversight Needs Further Enhancement*. GAO-07-883, July 31, 2007.

[6] For example, see EPA's Office of Inspector General Report, *Congressional Request on EPA Enforcement Resources and Accomplishments*, October 10, 2003, Rpt #- 2004-S-00001, http://www.epa.gov/oig/.

[7] The Environmental Council of the States (ECOS) is a national nonprofit (501(c)(6)), nonpartisan association of state and territorial environmental commissioners, established in December 1993. http://www.ecos.org.

[8] ECOS, *The Funding Gap*, The Journal of the Environmental Council of the States, Winter 2004. http://www.ecos.org/ section/publications.

[9] GAO, *Environmental Protection: EPA-State Enforcement Partnership Has Improved, but EPA's Oversight Needs Further Enhancement*. GAO-07-883, July 31, 2007.

[10] Title VI—Additional General Provisions Sec. 605 of H.R. 2643 (June 28, 2007, placed on the Calendar of the Senate). See also H.Amdt. 441 to H.R. 2643, offered and agreed to, June 26, 2007, inserted under Title VI—Additional General Provisions Sec. 601 (CR H.R. 7222). The FY2008 omnibus appropriations (P.L. 110-161) did not include this general provision (Joint Explanatory Statement Accompanying Division F of the Consolidated Appropriations Act for FY2008 (P.L. 110-161, H.R. 2764), as presented in the *Congressional Record*, December 17, 2007, pg. H16142).

[11] Conference Report 108-72 (p. 1563) accompanying the Consolidated Appropriations Act, 2005, P.L. 108-447); Conference Report 108-401 (p. 1126) accompanying the Consolidated Appropriations Act, 2004, P.L. 108-199; and Conference Report 108-10 (p. 1445) accompanying Consolidated Appropriations Resolution, 2003, P.L. 108-7.

[12] EPA's Office of Inspector General Report 2004-S-00001, *Congressional Request on EPA Enforcement Resources and Accomplishments*, October 10, 2003, http://www.epa.gov/oig/.

[13] See 33 U.S.C. §1319, 42 U.S.C. §7413, and 42 U.S.C. §300g-3.

[14] See CRS Report RL32240, *The Federal Rulemaking Process: An Overview*, by Curtis W. Copeland.

[15] For more information regarding EPA's current organizational structure for enforcement, see the agency's website at http://www.epa.gov/compliance

[16] See EPA, OECA, *Compliance and Enforcement Short Term Planning* website for individual annual fiscal year National Program Managers Guidance, and Memoranda of Understanding (MOAs), http://www.epa.gov/ compliance data/planning/shortterm.html.

[17] For more information regarding EPA's environmental enforcement priorities, and strategic planning, see EPA-OECA, *National Priorities for Enforcement and Compliance Assurance*, http://www.epa.gov/compliance planning/priorities/index.html.

[18] EPA's National Enforcement Investigations Center (NEIC) is located in Denver, Colorado. See http://www.epa.gov/ compliance/neic/index.html.

[19] EPA's National Enforcement Training Institute (NETI), http://www.netionline.com/.

20 EPA-OECA, *EPA FY2008 Compliance and Enforcement Annual Results Report, December 2008*, http://epa.gov/compliance/data/results/annual/index.html.

21 EPA-OECA discontinued reporting criminal referrals beginning with reporting in FY2005. EPA, *National Enforcement Trends FY2004—Criminal Enforcement*, http://www.epa.gov/ compliance nets04-criminal-enforcement.pdf

22 EPA's cases are typically handled by three of the division's 10 sections: the Environmental Crimes Section, the Environmental Enforcement Section, and the Environmental Defense Section (http://www.usdoj.gov/enrd/About_ENRD.html).

23 The term "delegated authority" has become the most commonly used when referring to EPA's authority to approve states' programs. Federal statutes more often use "primary enforcement responsibility," "primacy," "approved," or "authorized" states' responsibility.

24 See CRS Report RL30798, *Environmental Laws: Summaries of Major Statutes Administered by the Environmental Protection Agency (EPA)*, for references to sections of individual acts that provide state authority.

25 See footnote footnote 24, p. 48.

26 See EPA guidance for issuing federal inspector credentials to state/tribal governments to conduct civil inspections: http://www.epa.gov/compliance

27 See, CRS Report RL3 0030, *Clean Water Act: A Summary of the Law*, by Claudia Copeland.

28 The Environmental Council of the States (ECOS) has tracked delegated authority by state and statute; see http://www.ecos.org/section/states.

29 EPA-OECA, *Best Practices and Program Improvements Expected to Result from SRF*, September 12, 2007, http://www.epa.gov/compliance

30 EPA, FY 2009 Office of Enforcement and Compliance Assurance (OECA)National Program Manager (NPM) Guidance, revised June 2008, http://www.epa.gov/compliance

31 The Environmental Council of the States (ECOS) is a national nonprofit (501(c)(6)), nonpartisan association of state and territorial environmental commissioners.

32 See http://www.epa.gov/ocirpage/nepps/ for information regarding NEPPS.

33 Provisions of the Clean Water Act (CWA) under §309 are often cited as an example of legislation limiting EPA's authority to overfile. EPA's authority to enforce under this section is only limited when a state has commenced an "appropriate enforcement action" in response to and within 30 days of EPA's issuance of a notice of violation to the state (33 U.S.C. §1319(a)(1)); when the state has "commenced and is diligently prosecuting" an action under comparable state law; or when a penalty assessed under a state-issued final order has been paid, the violation will not be subject of a civil penalty action under §1319(d) §1321(b) or §1365 (33 U.S.C. §1319(g)(6)).

34 Ellen R. Zahren, *Overfiling Under Federalism: Federal Nipping at State Heels to Protect the Environment*, 49 Emory L. J. 373, 375 n.18 (2000); Joel A. Mintz, *Enforcement "Overfiling" in the Federal Courts: Some Thoughts on the Post-Harmon Cases*, 21 Virginia Envtl. L. J. 425, 427 (2003). Jeffrey G. Miller, *Theme & Variations in Statutory Preclusions Against Successive Environmental Enforcement Actions by EPA & Citizens, Part II Statutory Preclusions on EPA Enforcement*, 29 Harvard Envtl. L. Rev. 1, 3 (2005)

35 EPA-approved/authorized state programs generally do not apply in Indian country.

36 Some pollution control laws have been amended to clarify the role of tribal governments in the implementation of federal environmental programs. For example, from 1986 to 1990, Congress amended the Clean Water Act (33 U.S.C. § 1377(e)(2)), Safe Drinking Water Act (42 U.S.C. § 300j-11(b)(1)(B)), and Clean Air Act (42 U.S.C. §7601(d)(2)(B) to authorize EPA to treat Indian tribes in the same manner as states for purposes of program authorization.

37 For example, see Executive Order No. 13175 on Consultation and Coordination With Indian Tribal Governments, 65 *Federal Register* 67249 (November 9, 2000); Executive Memorandum on Government-to-Government Relations with Native American Tribal Governments, April 29, 1994.

38 Issued by President Ronald Reagan, the policy expanded the 1970 national Indian policy of self-determination for tribes, http://www.epa.gov/tribalportal/basicinfo/presidential-docs.html.

39 In conjunction with the 1983 overall federal policy statement, EPA consolidated existing agency statements into a single policy statement to ensure consistency. See *EPA Policy for the Administration of Environmental Programs on Indian Reservations*, http://www.epa.gov/superfund/community

40 EPA-OECA, National Priorities for Enforcement and Compliance Assurance, *Enforcement and Compliance Assurance Priority: Indian Country*, http://www.epa.gov/compliance

41 Although not strictly speaking "enforcement," citizens may also petition for review of agency actions under a program statute or the Administrative Procedure Act.

42 EPA-OECA, *National Report a Violation*, http://www.epa.gov/tips.

43 EPA-OECA, *Compliance and Enforcement Annual Results: Report a Violation*, http://www.epa.gov/compliance resources/reports/endofyear/eoy2006/sp-reportaviolation.html.

44 EPA-OIG, Limited Knowledge of the Universe of Regulated Entities Impedes EPA's Ability to Demonstrate Changes in Regulatory Compliance, 2005-P-00024, September 19, 2005 (http://www.epa.gov/oig/); and GAO,

Human Capital: Implementing an Effective Workforce Strategy Would Help EPA to Achieve Its Strategic Goals, GAO-01-812, pp. 24-25, July 2001, http://www.gao.gov/docsearch/repandtest.html.

[45] EPA-OECA, *OECA Regulatory Universe Identification Table. Internal EPA memorandum November 15, 2001, EPA-OIG, Limited Knowledge of the Universe of Regulated Entities Impedes EPA's Ability to Demonstrate Changes in Regulatory Compliance*, 2005-P-00024, September 19, 2005, http://www.epa.gov/oig/.

[46] ECOS, *State Environmental Contributions to Enforcement and Compliance:2000-2003*, June 2006, http://ecos.org/ section/publications.

[47] EPA, *National Emissions Inventories for the U.S.*, http://www.epa.gov/ttn/chief/index.html.

[48] Under §106 of the Clean Air Act, EPA has set National Ambient Air Quality Standards for six principal pollutants classified by the EPA as "criteria pollutants": sulfur dioxide (SO2), nitrogen dioxide (NO2), carbon monoxide (CO), ozone, lead, and particulate matter.

[49] Under §112 of the Clean Air Act, EPA is to establish technology-based emission standards, called "MACT" standards, for sources of 188 pollutants listed in the legislation, and to specify categories of sources subject to the emission standards.

[50] For more information regarding EPA's Sector Compliance Assistance and Sector Notebooks, see http://www.epa.gov/compliance

[51] Most federal environmental laws contain provisions that subject federal facilities to federal, state, and local requirements, and allow such facilities to be sued just as a nongovernmental entity. In addition, such provisions generally grant the President authority to exempt federal facilities from such requirements when in the "paramount interest" or (less commonly) the "national security interest" of the United States. See Clean Air Act (42 U.S.C. §7418), Clean Water Act (33 U.S.C. §1323), Resource Conservation and Recovery Act (42 U.S.C. §6961), and Safe Drinking Water Act (42 U.S.C. §300j-6). A more limited federal facility provision and presidential exemption is found in the Comprehensive Environmental Response, Compensation and Liability Act (42 U.S.C. §9620). The Toxic Substances Control Act nowhere expressly states that federal facilities are subject to the statute, but nonetheless does authorize presidential exemptions (15 U.S.C. §2621). For discussion of exemptions, particularly as they pertain to DOD, see CRS Report RS22149, *Exemptions from Environmental Law for the Department of Defense (DOD)*, by David M. Bearden.

[52] EPA-OECA, *Compliance and Enforcement Annual Results: FY2008 Federal Facilities*, http://www.epa.gov/ compliance/resources/reports/endofyear/eoy2008/2008-sp-fedfac.html.

[53] For examples of executive orders with directives addressing environmental management at federal facilities, see EPA's *Federal Facilities Sector Notebook: A Profile of Federal Facilities*, EPA 300-B-96-03, January 1996, pgs. 2-11 and 2-12, available at http://www.epa.gov/compliance

[54] EPA Federal Facilities Enforcement Office, *The State of Federal Facilities An Overview of Environmental Compliance at Federal Facilities FY 2003-2004*. EPA Document No: EPA 300-R-05-004. November 2005. http://cfpub.epa.gov/compliance

[55] Ibid.

[56] DOE accounted for $182 billion of the estimated environmental compliance costs. Nearly all of this amount was for the cleanup of nuclear weapons production sites. DOD accounted for $64 billion of the estimated costs. Most of this amount was for the cleanup of contamination from the past release of hazardous substances, and the cleanup of military training ranges where unexploded ordnance is present. The disposal of radioactive waste from nuclear-powered ships and submarines, and the disposal of chemical weapons stockpiles required by the Chemical Weapons Conventions, also constituted a sizeable portion of these costs. GAO, *Environmental Liabilities: Long-Term Fiscal Planning Hampered by Control Weaknesses and Uncertainties in the Federal Government's Estimates*, March 2006. GAO-06-427.

[57] 42 U.S.C. §6961.

[58] For persons who willfully or knowingly disregard the law.

[59] See the discussion later in this chapter on Supplemental Environmental Projects (SEPs).

[60] EPA-OECA, *EPA FY2008 Compliance and Enforcement Annual Results Report, December 2008*, http://epa.gov/ compliance/data/results/annual/index.html.

[61] Ibid.

[62] The CWA requires evaluations instead of investigations, which include reviewing reports, records, and operating logs; assessing air pollution control devices and operations; and observing visible emissions or conducting stack tests.

[63] EPA-OECA, *EPA FY2008 Compliance and Enforcement Annual Results Report, December 2008*, http://epa.gov/ compliance/data/results/annual/index.html.

[64] "State Environmental Agency Contributions to Enforcement and Compliance: 2000-2003," the Environmental Council of the States (ECOS), June 2006.

[65] EPA-OECA, *EPA FY2008 Compliance and Enforcement Annual Results Report, December 2008*, http://epa.gov/ compliance/data/results/annual/index.html.

[66] Administrative Procedure Act, 5 U.S.C. § 557.

[67] 5 U.S.C. §3105.

[68] Title 40 CFR Part 22 Subpart A § 22.4(b).

[69] Title 40 C.F.R. Part 22 Subpart A § 22.8.

[70] Title 40 C.F.R. § 1.25(e). EAB judges are Senior Executive Service ("SES")-level career Agency attorneys (U.S. EPA, *The Environmental Appeals Board Practice Manual, June 2004,* available at http://www.epa.gov/eab/pmanual.pdf).

[71] EPA, *Compliance and Enforcement Annual Results: FY2008 Criminal Enforcement,* http://www.epa.gov/compliance/resources/reports/endofyear/eoy2008/2008-sp-criminal.html.

[72] For examples of criminal cases for various fiscal years, see EPA's compliance and enforcement annual results website at http://www.epa.gov/compliance/ resources

[73] For a listing of EPA's Criminal Investigation Division Offices and associated Environmental Crimes Task Force Teams throughout the United States, see http://www.epa.gov/Compliance/ criminal/ intergovernmental/ environcrimes.html.

[74] For a historical perspective of EPA's criminal enforcement, see *EPA Review of the Office of Criminal Enforcement, Forensics, and Training,* November 2003, http://www.epa.gov/Compliance/ resources/

[75] Ibid.

[76] Based on information provided directly to CRS by EPA's Office of Congressional and Intergovernmental Relations, October 31, 2007, and March 26, 2008.

[77] Ibid.

[78] In the context of this chapter, "sanctions" refer to adverse consequences imposed in response to noncompliance.

[79] 2 CFR Part 180, 2 CFR Part 1532, and Executive Order 12549. Debarment is also authorized following criminal conviction under Clean Water Act (§508) and Clean Air Act (§306). See information regarding EPA's suspension and debarment program at http://www.epa.gov/ogd/sdd/debarment.htm.

[80] Under the Miscellaneous Receipts Act (31 U.S.C. §3302(b)), all court, or administratively imposed penalties must be paid by the government official receiving the monies to the U.S. Treasury.

[81] See, for example, 33 U.S.C. §1321(f)(5).

[82] SDWA, 42 U.S.C. §300g-2(a)(6), added by SDWA Amendments of 1996 (P.L. 104-182).

[83] See CRS Report RL3 1243, *Safe Drinking Water Act (SDWA): A Summary of the Act and Its Major Requirements,* by Mary Tiemann.

[84] EPA guidance for the "Division of Penalties with State and Local Governments," memorandum from Courtney M. Price, Assistant Administrator for Enforcement and Compliance Monitoring, October 30, 1985.

[85] EPA, Office of Enforcement and Compliance Assurance, *FY2006 OECA Accomplishments Report,* Spring 2007, EPA-300-R-07-001, http://cfpub.epa.gov/ compliance

[86] For example, EPA's revised *Consolidated Rules of Practice* ("CROP") (64 FR 40138, July 23, 1999) contains procedural rules for the administrative assessment of civil penalties, issuance of compliance or corrective action orders, and the revocation, termination or suspension of permits, under most environmental statutes. See EPA's Web page, *Civil Penalty Policies,* at http://cfpub.epa.gov/compliance/ resources

[87] Additionally, see EPA, *Guidance on Calculating the Economic Benefit of Noncompliance by Federal Agencies,* issued on February 13, 2006, Memorandum from Granta Y. Nakayama, Assistant Administrator, Office of Enforcement Compliance and Assurance, http://www.epa.gov/Compliance/resources enforcement/cleanup/guid-econ-ben-noncomp-2-13-06.pdf.

[88] Prepared by Robert Meltz, Legislative Attorney, American Law Division.

[89] *U.S. Dept. Of Energy v. Ohio,* 503 U.S. 607 (1992).

[90] EPA-OECA, *Compliance and Enforcement Annual Results: FY2008 Federal Facilities,* http://www.epa.gov/compliance

[91] For more information regarding SEPs, see http://www.epa.gov/compliance

[92] Policy and guidance documents related to EPA's Supplemental Environmental Projects Policy are available at http://cfpub.epa.gov/compliance

[93] EPA-OECA, *Compliance and Enforcement Annual Results: FY2008 Numbers at a Glance,* http://www.epa.gov/compliance

[94] The Environmental Council of States (ECOS), *State Environmental Agency Contributions to Enforcement and Compliance 2000-2003,* June 2006.

[95] House Committee on Energy and Commerce, Subcommittee on Environment and Hazardous Materials: H.R. 1103, the Environmental Justice Act of 2007, and H.R. 1055, the Toxic Right-To-Know Protection Act, Legislative Hearing—*Environmental Justice and the Toxics Release Inventory Reporting Program: Communities Have a Right to Know,* October 4, 2007. Senate Environment and Public Works Committee, Subcommittee on Superfund and Environmental Health Hearing: *Oversight of the EPA's Environmental Justice Programs,* July 25, 2007.

[96] Executive Order 12898, 49 *Federal Register* 7629, February 16, 1994, http://www.archives.gov/federal-register/executive-orders/1994.html.

[97] EPA's Office of Civil Rights (OCR) is responsible for the agency's administration of Title VI of the Civil Rights Act of 1964, including processing and investigating administrative complaints under implementing regulations (40 C.F.R. Part 7) prohibiting EPA-funded permitting agencies from "... permitting actions that are

intentionally discriminatory or have a discriminatory effect based on race, color, or national origin." See http://www.epa.gov/civilrights/t6home.htm.

[98] EPA reports that, since 1994, more than $31 million in funding has been provided to more than 1,100 community- based organizations. For more information regarding EPA's Environmental Justice Program activities, see http://www.epa.gov/compliance

[99] EPA-OECA, "Environmental Justice Strategic Enforcement Assessment Tool (EJSEAT)," http://www.epa.gov/compliance/resources/policies/ej/index.html.

[100] EPA-OECA, Office of Environmental Justice, EPA 300-R-04-002, http://www.epa.gov/compliance policies/ej/ej-toolkit.pdf.

[101] EPA-OECA, *Environmental Justice Geographic Assessment Tool*, http://www.epa.gov/compliance environmentaljustice/assessment.html.

[102] ECOS, State Agency Contributions to Enforcement and Compliance, ECOS 01-004, April 2001.

[103] For more information regarding EPA's compliance assistance programs, see http://www.epa.gov/compliance/assistance/index.html.

[104] EPA-OECA, http://www.epa.gov/Compliance/assistance/planning

[105] For more information regarding the National Compliance Assistance Centers, see http://www.epa.gov/compliance assistance/centers/index.html or http://www.assistancecenters.net/.

[106] EPA National Compliance Assistance Clearinghouse, available at http://134.67.99.39/clearinghouse/index.cfm?TopicID=C: 10:600:.

[107] For more information regarding EPA's incentive programs and initiatives, see http://www.epa.gov/compliance incentives/index.html.

[108] 72 *Federal Register* 27116, May 14, 2007.

[109] EPA-OECA, *Compliance and Enforcement Short Term Planning*, http://www.epa. gov/compliance

[110] EPA-OIG: *Performance Track Could Improve Program Design and Management to Ensure Value*, Report No. 2007-P-00013, March 29, 2007. GAO: Environmental Protection: *Challenges Facing EPA's Efforts to Reinvent Environmental Regulation*, RCED-97-155, July 2, 1997.

[111] Title VI—Additional General Provisions, Sec. 605 of H.R. 2643 as placed on the Calendar of the Senate (June 28, 2007).

[112] Joint Explanatory Statement Accompanying Division F of the Consolidated Appropriations Act for FY2008 (P.L. 110-161, H.R. 2764), as presented in the Congressional Record, December 17, 2007 (pg. H16142).

[113] EPA's appropriations are within the jurisdiction of the Interior, Environment, and Related Agencies appropriations subcommittees.

[114] DOJ, Justice Management Division, Budget Staff, *Department of Justice FY2009 Budget Congressional Submission* http://www.usdoj.gov/jmd/2009justification/, and information provided by ENRD in written communication to CRS. See also, *Budget Trend Data 1975 Through the President's 2003 Request to the Congress*, pp. 55-61, Spring 2002, http://www.usdoj.gov/jmd/budgetsummary/btd/1975_2002/btd02tocpg.htm.

[115] All dollar amounts are adjusted to a 2003 basis for inflation. *ECOS Budget Survey: Budgets are bruised, but Still Strong*, R. Steven Brown and Michael J. Kiefer, Summer 2003 *ECOStates* (http://www.ecos. org/section /states/ spending).

[116] EPA, *FY2006 OECA Accomplishment Report*, EPA-300-R-07-001, Spring 2007, http://cfpub.epa.gov/compliance resources/reports/accomplishment/details.cfm.

[117] For a more complete list and descriptions of EPA's enforcement/compliance databases, see http://www.epa.gov/compliance/data/systems/index.html.

[118] See "EPA Results and Reports" at http://cfpub.epa.gov/compliance

[119] EPA-OIG: Overcoming Obstacles to Measuring Compliance: Practices in Selected Federal Agencies, Report No. 2007-P-00027, June 20, 2007; EPA Performance Measures Do Not Effectively Track Compliance Outcomes, Report No. 2006-P-00006; Congressional Request on Updating Fiscal 2003 EPA Enforcement Resources and Accomplishments, Report 2004-S-00002 http://www.epa.gov/oig/. GAO: Environmental Enforcement: EPA Needs to Improve the Accuracy and Transparency of Measures Used to Report on Program Effectiveness, GAO-08-1111R, September 18, 2008 http://gao.gov.

[120] Several states have provided direct links to related websites, which EPA posts on the ECHO website at http://www.epa-echo.gov/echo/more_state_data.html.

In: Enforcing Federal Pollution Control Laws
Editor: Norbert Forgács

ISBN: 978-1-60876-082-4
© 2010 Nova Science Publishers, Inc.

Chapter 6

OECA FY 2008 ACCOMPLISHMENTS REPORT: PROTECTING PUBLIC HEALTH AND ENVIRONMENT[*]

United States Environmental Protection Agency
Office of Enforcement and Compliance Assurance

MESSAGE FROM THE ASSISTANT ADMINISTRATOR

If someone had told me three years ago when I took the helm of EPA's Office of Enforcement and Compliance Assurance (OECA) that we would be protecting our nation's air, water, and land at a pace never before seen in EPA's history, I would have expressed skepticism. Today that skepticism is replaced with pride. OECA's accomplishments for Fiscal Year (FY) 2008 are exceptional—in several instances reaching record levels and even surpassing the combined historic records of previous years.

The strength of EPA's enforcement program is illustrated by an unprecedented run of record results. EPA holds polluters accountable. In FY 2008, EPA concluded civil and criminal enforcement actions requiring polluters to spend an estimated $11.8 billion, an agency record, on pollution controls, cleanup and environmental projects. This exceeds the

---

[*] This is an edited, reformatted and augmented version of a U. S. Environmental Protection Agency publication dated December 2008.

FY 2007 amount by approximately $800 million.* This means that each workday OECA was securing agreements from violators to invest an estimated $47 million to achieve compliance. The combined total for the last five years is an estimated $45 billion ($5.5, $11.3, $5.4, $11.0, and $11.8 billion, respectively)—exceeding EPA's total annual budget over the same period.

After all the complying actions for FY 2008 cases are completed, EPA estimates that 3.9 billion pounds of pollution will be reduced or removed annually from the environment, the highest amount since FY 1999. In the last five years EPA's record for estimated pollution reductions stood at 1.1 billion pounds for FY 2005. The estimated pollutant reductions resulting from FY 2008 enforcement actions exceed FY 2005 by almost four times. The FY 2008 estimate also exceeds the combined results obtained during FY 2004–2007 by nearly 100 million pounds.

Nearly half of this year's pollution reductions are the result of an enforcement action taken against American Electric Power, one of the largest environmental settlements of all time. EPA, along with our partners at the U.S. Department of Justice, and the States of New York, Connecticut, New Jersey, Vermont, New Hampshire, Maryland, Rhode Island, and the Commonwealth of Massachusetts, negotiated this historic settlement which will save an estimated $32 billion in health costs per year.

In addition to achieving substantial pollutant reductions, FY 2008 settlements included significant penalties for violations of environmental requirements. Penalties assessed by EPA play an important role in deterring potential polluters from violating environmental laws and regulations. EPA assessed approximately $127 million in civil penalties and courts sentenced defendants to pay $64 million in criminal fines.

In January 2008, EPA secured a $20 million civil penalty from Massey Energy, the largest coal producer in Central Appalachia. This penalty is the largest in EPA's history levied against a company for wastewater discharge permit violations. The Massey settlement will not only improve fish habitat, but will also reduce downstream flooding, benefiting a number of poor, rural communities in Kentucky and West Virginia.

In FY 2008, EPA obtained commitments from responsible parties to invest nearly $1.6 billion for investigations and cleanup of Superfund sites. This is the highest total in seven years, and the fifth highest total in the history of the Superfund enforcement program.

Through our compliance assistance activities, EPA reached over 2.5 million entities. EPA's compliance assistance programs include Web sites and guidance that provide detailed information to millions of regulated entities, helping them understand and meet their environmental obligations. Last year, EPA launched a new compliance assistance center (www.campuserc.org) that provides comprehensive compliance assistance and pollution prevention information for regulated activities at nearly 4,200 colleges and universities.

Since its inception two years ago, our *Tips and Complaints* Web site has received over 18,000 tips and resulted in opening 19 criminal cases. Last year, citizen tips resulted in two criminal convictions.

In FY 2008, OECA's oversight of import and export hazardous waste notices prevented the environmentally unsound importation of 97,000 tons of hazardous waste. OECA also developed a framework between the United States and China to establish training and programs on exported and imported products to protect human health and the environment.

We work together with our partners at the U.S. Department of Justice and state governments to achieve these results, and are proud of what we have accomplished. The commitment of our staff and government partners was paramount in achieving our historic results. These results will have lasting benefits for all people.

Sincerely,

Granta Y. Nakayama

EPA ASSISTANT ADMINISTRATOR FOR
ENFORCEMENT AND COMPLIANCE ASSURANCE

OECA's Mission

OECA is responsible for monitoring compliance with environmental laws, providing compliance information and assistance to the regulated community, and taking civil or criminal enforcement action when needed. OECA's goal is to ensure that the environmental and public health benefits that are promised by our nation's environmental laws are realized. The diagram below illustrates how these activities work together to accomplish that goal.

> "OECA's mission is to improve the environment and protect public health by ensuring compliance with the nation's environmental laws."
> *EPA Strategic Plan*

OECA's responsibilities also include two other vital EPA programs: environmental justice and National Environmental Policy Act (NEPA) reviews. See the organizational chart at Appendix A.

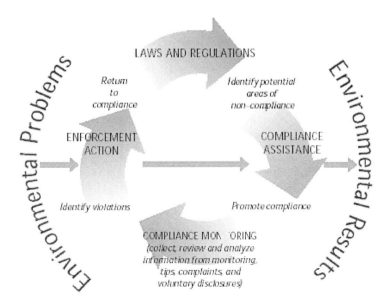

Enforcement & Compliance Lifecycle

OECA's Workforce and Partners

In FY 2008, OECA had a workforce of about 3,300 environmental professionals. Two-thirds of these employees are located in EPA's ten Regional offices, where they work closely with our state partners to monitor and enforce compliance with the Nation's environmental laws. OECA also works closely with the U.S. Department of Justice, which represents EPA in federal court enforcement actions, and with other federal agencies on their NEPA decisions.

About This chapter

This chapter highlights the accomplishments of OECA's enforcement, compliance and other programs in FY 2008. The report explains key priorities and strategies, long-term trends, and the results that OECA's programs have obtained for the public.

We encourage you to visit our Web site at www.epa.gov/compliance for more information about OECA and its programs, our enforcement cases and annual results.

FY 2008 RESULTS AT A GLANCE

In FY 2008, the enforcement actions concluded will reduce pollutant emissions to air, water and land by an estimated 3.9 billion pounds per year, when the pollution controls and other measures required by these actions are installed and operational. This is more than four times the level of pollutant reductions accomplished in FY 2007, and nearly equals the four prior years combined, as shown by the table below.

> "Strong enforcement is a key to ensuring that the promise of our environmental statutes is matched by the environmental reality."
>
> *Lynn Buhl, Regional Administrator EPA Region 5*

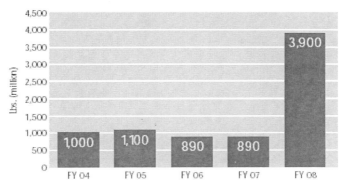

FY 2008 Data Source: Integrated Compliance Information System (ICIS), October 11, 2008; data source for previous fiscal years: ICIS.

Estimated Pollutant Reduction Commitments

In addition, EPA obtained enforcement commitments from parties responsible for managing hazardous waste to treat, minimize or properly dispose of an estimated 6.5 billion pounds of hazardous waste.

These pollutant reductions will result from legally enforceable commitments by violators who were not in compliance with the law to invest a total of over $11.8 billion, the highest amount on record, in pollution controls, cleanup, and environmentally beneficial projects. (See Appendix B for a detailed summary of our enforcement and compliance results.)

FY 2008 Other Highlights

Civil Penalties. EPA obtained nearly $127 million in civil penalties through civil judicial and administrative enforcement actions this year. This represents an increase of $55 million over FY 2007.

Superfund Enforcement

EPA maintained very vigorous Superfund enforcement activity in FY 2008, ensuring that polluters, rather than the public, pay for cleanups of Superfund sites. We obtained commitments from responsible parties to invest approximately $1.6 billion for investigation and cleanup of contaminated sites. This will result in the cleanup of an estimated 100 million cubic yards of contaminated soil, an all-time record for the Superfund enforcement program, and about 255 million cubic yards of contaminated ground water. We also obtained reimbursement from responsible parties of $232 million of EPA's past costs for Superfund site investigations and cleanups.

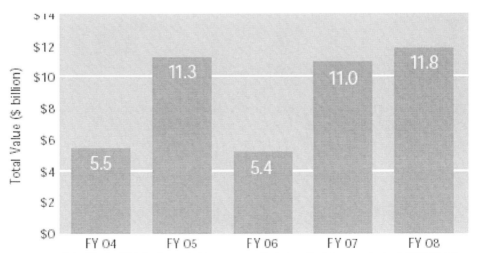

Estimated Investments in Pollution Control and Cleanup plus Environmentally Beneficial Projects (Inflation Adjusted to FY 2008 Dollars)

142   United States Environmental Protection Agency

Criminal Enforcement

EPA's criminal enforcement program obtained sentences totaling 57 years of incarceration, $64 million in fines and restitution, and $12 million in court-ordered environmental projects. The relief obtained in criminal cases will result in pollutant reductions totaling 1.6 million pounds.

Compliance Monitoring and Assistance

EPA conducted 20,000 facility inspections and evaluations in FY 2008, maintaining a strong presence that deters and detects non-compliance with the nation's environmental laws. Many more inspections were conducted by our state partners across the country, vastly expanding our ability to deter and detect potential violations. EPA brought compliance assistance to a wide audience of over 2.5 million through presentations, workshops, onsite visits and responses to inquiries, as well as indirect outreach via mailings and internet-based assistance. EPA's compliance assistance resources help small and medium-sized businesses meet their compliance responsibilities.

GETTING THE WORD OUT: EFFECTIVE COMPLIANCE ASSISTANCE

EPA's compliance assistance programs provide information to millions of regulated entities, particularly small businesses, to help them understand and meet their environmental obligations. This information lets regulated entities know of their legal obligations under federal environmental laws. Compliance assistance resources include comprehensive Web sites, compliance guides, and training materials aimed at specific business communities or industry sectors. Also, onsite compliance assistance and information is sometimes provided by EPA inspectors.

Web-based Compliance Assistance Centers

EPA provides effective and efficient compliance information to regulated entities, primarily small businesses, through 16 Web-based compliance assistance centers. The centers assist users by providing compliance tools and contacts for over 20 topics, including federal requirements for control of contaminated stormwater, air and hazardous waste, lead and mercury. In addition, the centers provide easy access to state-specific regulations and compliance resources.

The regulated community relies heavily upon the compliance assistance center Web sites. During FY 2008, EPA reached more than 2.2 million entities through online compliance assistance activities.

The Web centers reach a much larger audience than other methods of compliance assistance, and have provided an increasingly large proportion of EPA's compliance assistance over the past five years.

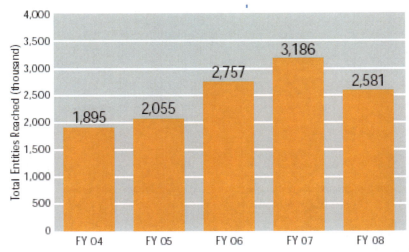

FY 2008 Data Sources: Integrated Compliance Information System (ICIS), October 11, 2008 and online usage report; data source for previous fiscal years: ICIS and on-line usage reports.

Entities Reached with EPA Compliance Assistance

SECTORS SERVED BY COMPLIANCE ASSISTANCE CENTERS

- Agriculture
- Auto Recycling
- Auto Repair
- Border Compliance
- Chemical
- Colleges/Universities
- Construction
- Federal Facilities
- Healthcare
- Local Government
- Metal Finishing
- Paints and Coatings
- Printed Wiring Board
- Printing
- Transportation
- Tribal Governments

Visit: www.epa.gov/compliance/assistance/centers

In FY 2008, EPA launched a new online center, the Campus Environmental Resource Center (www.campuserc.org), to provide comprehensive environmental compliance assistance and pollution prevention information. This Web center assists colleges and universities to identify the campus areas to which environmental requirements apply, e.g., laboratories and hazardous waste disposal, and to ensure that they are in compliance with the law.

This year EPA also enhanced an online center (www.bordercenter.org) to address compliance issues at our international borders. This Web site provides information on hazardous waste transport issues across the Mexican border, municipal solid waste transport across the Canadian border, and requirements applicable to small non-road engines, in response to an increase in imports of polluting engines from Asia.

In recognition of the success of EPA's efforts, EPA received a special recognition award from the Small Business Administration for its "*extraordinary responsiveness and service to small businesses regarding compliance and enforcement issues.*" The 2007 National Small Business Ombudsman's Report to Congress recognized EPA's Web compliance centers as "*practical tools that assist small businesses by providing comprehensive, easily accessible federal and state compliance and pollution prevention information.*"

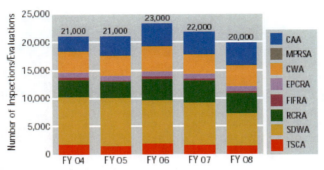

Note: Statutes in legend are presented in same order as in stacked bars on left. FY 2008 Data Source: Integrated Compliance Information System (ICIS), legacy databases, and manual reporting, October 11, 2008. Data source for previous fiscal years: ICIS, legacy databases, and manual reporting.

Number of Inspections/Evaluations Conducted by EPA

THE ENVIRONMENTAL COP IS ON THE BEAT: COMPLIANCE MONITORING

One of EPA's key responsibilities is to monitor compliance with the nation's environmental laws, to deter and detect violations. OECA monitors compliance through facility inspections by regional, state and tribal inspectors, as well as by reviewing the self-monitoring reports that are submitted by regulated entities for many environmental programs. OECA maintains a large national compliance database, which collects the results of inspections and self-monitoring reports. We also make compliance information available to the public through our *Enforcement and Compliance History Online* (ECHO) Web site at www.epa-echo.gov/echo.

> "All facilities that produce hazardous pollutants must carefully adhere to all provisions of EPA's requirements to ensure that we are taking every necessary step to protect human health and our environment."
>
> *Robert W. Varney, Regional Administrator EPA Region 1*

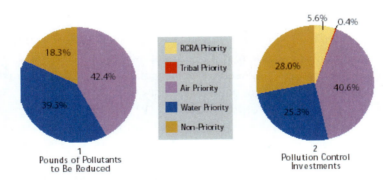

FY 2008 National Priority Contributions

In FY 2008, EPA conducted approximately 20,000 inspections, and 222 civil investigations (complex, in-depth examinations). In addition, our tribal partners, using federal credentials, conducted 334 inspections to monitor compliance with environmental laws in Indian Country. Many more inspections for compliance with national and state environmental laws were conducted by state inspectors.

DELIVERING ENVIRONMENTAL RESULTS: CIVIL ENFORCEMENT BREAKS RECORDS

In FY 2008, EPA's concluded enforcement actions will reduce pollutant emissions to air, water and land by an estimated 3.9 billion pounds per year when the pollution controls and other measures required by these actions are installed and operational.

These pollution reductions will result from legally enforceable commitments by violators to invest an estimated $11.8 billion, the highest amount on record, on installing pollution controls, cleanup and environmental projects.

Focus on National Enforcement Priorities Brings Results

EPA secured these record commitments by focusing on specific environmental programs and sectors that were selected as national priorities for enforcement attention. OECA achieved nearly 82 percent of the FY 2008 pollution reductions and 66 percent of the pollution control investments in our high-priority areas.

OECA selected the national priority areas by reviewing national data and compliance information, and soliciting input from our state partners and the public. This review identified areas of significant non-compliance with the nation's environmental laws across the country that resulted in substantial amounts of illegal pollution. OECA conducts this review every three years. EPA focused on the following national priorities during FY 2008:

Clean Air Act Enforcement Priorities

Clean Air Act/Prevention of Significant Deterioration and New Source Review

The New Source Review (NSR) and Prevention of Significant Deterioration (PSD) requirements of the Clean Air Act (CAA) require facilities in certain sectors to install state-of-the-art pollution controls when they are newly constructed or significantly modified. Failure to comply with these requirements in some sectors has led to illegal emissions of thousands of tons of pollutants, including sulfur dioxide (SO_2), nitrogen oxides (NO_X), volatile organic compounds (VOC), and particulate matter (PM). OECA's priority enforcement efforts have focused on coal-fired power plants, and glass, acid, and cement manufacturers.

Clean Air Act/Air Toxics

This priority focuses on enforcing compliance with the Maximum Achievable Control Technology (MACT) standards for control of toxic air pollutants from sources that emit hazardous air pollutants.

Clean Water Act Enforcement Priorities

OECA ensures compliance with Clean Water Act (CWA) requirements by addressing four environmental challenges that are exacerbated by wet weather. Wet weather discharges contain bacteria, pathogens, and other pollutants that can cause illnesses in humans, and lead to water quality impairment.

- Combined Sewer Overflows (CSOs): Combined sewer systems are designed to collect rainwater runoff, domestic sewage, and industrial wastewater in the same pipe. During periods of rainfall or snow melt, the wastewater volume in a combined sewer system can exceed the capacity of the system or treatment plant, leading to discharge of pollutants into waterways.

- Sanitary Sewer Overflows (SSOs): Sanitary sewers are designed to carry sewage only, but these sewers also can overflow when the system's capacity or operation and maintenance is inadequate. This can lead to the discharge of bacteria, pathogens, nutrients, untreated industrial wastes, toxic pollutants, such as oil and pesticides, and wastewater solids and debris into waterways.

- Stormwater: Stormwater runoff from urban areas, industrial areas, and construction sites can include a variety of pollutants, such as sediment, bacteria, organic nutrients, hydrocarbons, metals, oil, and grease. Violations of requirements for control of stormwater runoff can lead to discharge of these contaminants into waterways.

- Concentrated Animal Feeding Operations (CAFOs): CAFOs generate a large volume of animal waste in concentrated areas. When requirements for control of this waste are not met, the waste can contaminate surface and ground waters.

Resource Conservation and Recovery Act Enforcement Priorities

Mineral Processing Facilities

Mineral processing facilities are often extremely large facilities which produce a substantial amount of hazardous waste containing metals and often water with low pH. Over the past decade, EPA has found that many of the facilities that manage these wastes have contaminated ground water, surface water and soil either through failure to comply with state or federal environmental requirements or legally permissible waste management practices. Large-scale mineral processing and mining operations often severely affect water supplies and wildlife and create environmental damage. Some facilities are located in populated areas,

OECA FY 2008 Accomplishments Report 147

making health risks a significant concern for EPA. This enforcement priority seeks to ensure that these facilities are complying with requirements for the handling and disposal of hazardous waste.

Financial Responsibility

Hazardous waste facility operators are required to maintain adequate funding for facility closure, including ensuring that any spills or leaks are cleaned up. The funds provide for the ability to clean up hazardous materials so they do not contaminate soils, ground water, surface waters or the air. Having the financial resources to perform closure and cleanup are an important part of protecting human health and the environment from solvents, dioxins, oils, heavy metals, polychlorinated biphenyls (PCBs), and other dangerous pollutants. OECA has been giving priority attention to assuring that these vital financial protections are in place.

Indian Country Enforcement and Compliance Priority

Federally-recognized Indian tribes are often faced with significant human health and environmental problems associated with drinking water supplies, solid waste disposal, and environmental risks in Indian schools (e.g., asbestos, lead paint). For the thousands of tribal members dependent on approximately 800 public drinking water systems in Indian country, including those providing drinking water to schools, violations of health-based standards can result in serious illness. Illegal dumping of solid waste and hazardous waste poses significant threats to soil and ground water. Uncontrolled dumps may catch on fire releasing particulate matter and other pollutants into the air and ecosystem, and discarded pesticides and other chemicals may leach into ground water or run off into surface water. OECA and the EPA Regions are working to build the tribes' capacity to monitor and address these problems, as well as taking appropriate enforcement action to correct problems that occur in Indian Country.

Public Health Benefits

OECA's focus on air enforcement yields substantial benefits for the environment and public health. Air pollution threatens human health by causing serious respiratory problems and exacerbating childhood asthma.

FY 2008 AIR ENFORCEMENT CASES YIELD HUMAN HEALTH BENEFITS

EPA's 10 largest enforcement actions for stationary source Clean Air Act violations obtained commitments by companies to reduce their emissions of sulfur oxides (SO_X), nitrogen oxides (NO_X) and particulate matter (PM). The annual human health benefits from these reductions in SO_X, NO_X, and PM are estimated at \$35 billion. These health benefits include:

- Approximately 4,000 avoided premature deaths in people with heart or lung disease;
- Over 2,000 fewer emergency room visits for diseases such as asthma and respiratory failure;
- About 6,000 fewer cases of chronic bronchitis and acute bronchitis;
- About 4,000 fewer nonfatal heart attacks;
- Over 30,000 fewer cases of upper aggravated asthma;
- Over 50,000 fewer cases of upper and lower respiratory symptoms; and
- Over 200,000 fewer days when people would miss work or school.

Data Source of Pollutant Reduction: Integrated Compliance Information System (ICIS), October 11, 2008. Benefit Estimate: The estimate of benefits of reducing $PM_{2.5}$ and its precursors (SO_X and NO_X) was generated by Office of Air Quality Planning and Standards Organization (OAQPS).

Enforcement Case Highlights

The following examples reflect our FY 2008 enforcement agreements involving coal-fired electric power utilities, construction sites, mineral processors, and wastewater discharge permit holders.

Coal-fired Power Plants

Coal-fired power plants release SO_2, NOX, and PM which cause respiratory problems and contribute to childhood asthma, acid rain, smog, and haze. In one of the largest cases in EPA history, American Electric Power will cut an estimated 1.6 billion pounds of air pollution from its coal-fired power plants. The company will also pay a \$15 million penalty and spend \$60 million on projects to mitigate the adverse effects of past emissions.

Stormwater

Without onsite pollution controls, construction site runoff can flow directly to the nearest waterway and degrade water quality. Runoff contains pollutants such as concrete washout, paint, used oil, pesticides, solvents and other debris.

Four of the nation's largest home builders will pay more than \$4 million to prevent an estimated 1 .2 billion pounds of sediment from polluting our nation's waterways each year. The builders—KB Home, Centex, Pulte and Richmond—will implement comprehensive, company-wide programs to improve compliance. The builders must develop improved pollution prevention plans, increase site inspections, promptly correct problems, and ensure construction site staff are properly trained.

Mineral Processing

In FY 2008, EPA issued an order to Agrifos Fertilizer Inc. and ExxonMobil to address wastewater management and prevent future imminent and substantial endangerment to human health and the environment. In August 2007, a retaining wall at Agrifos' Pasadena, Texas mineral processing facility failed, releasing more than 50 million gallons of acidic wastewater into local waters causing the death of thousands of fish. The companies are required to take specific steps to properly treat and dispose of 1.75 billion pounds of hazardous waste per year.

Wastewater Discharge

Massey Energy, the largest coal producer in central Appalachia, will pay a $20 million civil penalty in a corporate-wide settlement to resolve CWA violations at coal mines in West Virginia and Kentucky. This is the largest civil penalty in EPA's history levied against a company for wastewater discharge permit violations. Massey will take measures to prevent an estimated 380 million pounds of pollutants from entering the water; invest approximately $10 million to develop a comprehensive system to prevent future violations; and set aside 200 acres of riverfront land in West Virginia for conservation purposes.

FY 2007 and 2008 Enforcement and Compliance Annual Results Priority Air, Water, Land & Financial Assurance Problems

Priority	Estimated Pollutants to be Reduced (millions of pounds)		Estimated Investments in Pollution Control (millions of dollars)	
	Priority Air Pollution Problems			
	FY 2007	FY 2008	FY 2007	FY 2008
NSR/PSD	426 M	1,654 M	$2,521 M	$4,790 M
Air Toxics	0.8 M	0.09 M	$11 M	$7 M
Total Air	426.8 M	1,654 M	$ 2,532 M	$4,787 M
	Priority Wet Weather Pollution Problems			
CSO/SSO	45 M	173 M	$3,635 M	$2,909 M
CAFO	15 M	32 M	$30 M	$10 M
Stormwater	118 M	1,329 M	$9 M	$68 M
Total Wet Weather	178 M	1,534 M	$3,674 M	$2,986 M
	Priority Land Pollution Problems			
Mineral Processing	NC*	1,751 M	$59M	$217 M
	Estimated Pounds of Hazardous Waste Treated, Minimized, or Properly Disposed of (millions of pounds)		Estimated Investments in Pollution Control (millions of dollars)	
Mineral Processing	NC*	1,751M	$60 M	$217 M
			Estimated Value of Financial Assurance Restored (millions of dollars)	
Financial Responsibility	NA*	NA*	NC*	$134 M

Note: All prior FY dollar figures in this chapter are adjusted to reflect the current value in FY 2008 dollars based on the monthly rate of inflation as determined by the U.S. Department of Labor Consumer Price Index for All Urban Consumers.

* NA = not applicable; NC = no data collected

Note: All prior FY dollar figures in this chapter are adjusted to reflect the current value in FY 2008 dollars based on the monthly rate of inflation as determined by the U.S. Department of Labor Consumer Price Index for All Urban Consumers. FY 2008 Data Source: Integrated Compliance Information System (ICIS), October 11, 2008; data source for previous fiscal years: ICIS.

Civil Penalties Assessed (Inflation Adjusted to FY 2008 Dollars)

Civil penalties play a significant role in deterring potential violators and "leveling the playing field" for those who comply with environmental laws. In FY 2008, EPA assessed about $127 million in civil penalties against defendants—nearly $50 million more than FY 2007.

Compliance Incentives for Proactive Behavior: EPA's Audit Policy and eDisclosure System

EPA provides incentives to companies that voluntarily discover, promptly disclose, correct, and prevent future environmental violations through the Audit Policy. EPA may reduce or waive penalties for violations if the facility meets the conditions of the policy. EPA will not waive or reduce penalties for repeat violations, or violations that resulted in serious actual harm.

The Audit Policy has yielded great results. Since 1995 more than 3,500 companies have disclosed and resolved violations at nearly 10,000 facilities under the policy. FY 2008 marks the highest total of facilities that disclosed violations in a single year—2,294 facilities.

Recognizing the success of the program, EPA decided to maximize results by taking the Audit Policy in some new directions. In FY 2008, EPA launched a new approach that offers incentives to new owners of facilities who correct environmental violations at recently-acquired facilities. Under the new approach, new owners may be eligible for reduced penalties. The new approach encourages owners of recently-acquired facilities to come forward, make a clean start by addressing environmental noncompliance, and promptly make changes to ensure they stay in compliance.

EPA made additional changes to streamline the process for everyone. Now, regulated entities can submit self-disclosures online through a new Web-based system, "eDisclosure." The new system allows facilities to submit their information securely on EPA's Web site, and should reduce transaction costs by ensuring that each disclosure contains complete information. The eDisclosure Web site can be found at www.epa.gov/compliance auditing/edisclosure.html.

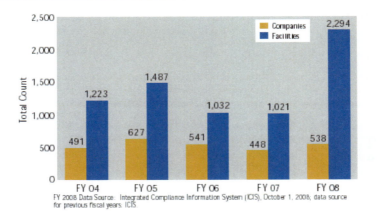

EPA Voluntary Disclosure Programs Voluntary Disclosures Initiated

ENVIRONMENTAL CRIME DOES NOT PAY

The mission of OECA's Office of Criminal Enforcement, Forensics and Training is to punish and deter serious environmental offenses. OECA's special agents, supported by forensic specialists at the National Enforcement Investigations Center, investigate allegations of criminal environmental violations, and work with criminal prosecutors at the U.S. Department of Justice to prosecute violators. Criminal enforcement actions are brought to address criminal violations of federal environmental statutes, as well as associated violations of the U.S. Criminal Code such as conspiracy, making false statements to investigators, interfering with an investigation, and mail fraud. Most of the environmental crimes that EPA pursues involve "knowing violations" of the law, which are classified as felonies.

> "Submitting false information in order to mislead authorities is illegal and will not be tolerated. The Justice Department will continue to work cooperatively with EPA and other law enforcement agencies to ensure the public's safety and protect our natural resources."
> *Ronald J. Tenpas, Assistant Attorney General for the U.S. Department of Justice's Environment and Natural Resources Division*

Criminal enforcement is the federal government's strongest sanction, with the possibility of incarceration of individuals, as well as significant monetary fines and restitution. The overall activities and results of EPA's criminal enforcement actions during FY 2008 are shown below.

EPA's criminal enforcement program addresses all of the environmental statutes and it uses a strategic approach to identify cases with significant environmental and human health impact, cases which enhance deterrence, and cases which advance EPA's enforcement priorities. Thirty four national enforcement priority criminal cases were opened in FY 2008, with six resulting in formal charges being filed during the year.

A prosecution that advances EPA's water enforcement priority was brought against Archer Daniels Midland (ADM) Company's Chattanooga, Tennessee facility, which manufactures high-quality paper products from raw cotton. The company lacked equipment needed to contain spills and other releases. ADM pled guilty to negligently discharging pollutants from the plant into Chattanooga Creek, a tributary of the Tennessee River, and was sentenced to pay a $100,000 criminal fine and another $100,000 in restitution to three environmental agencies and associations.

EPA investigators give priority to cases involving actual and threatened harm to human health or the environment. In a case against British Petroleum Exploration (Alaska), Inc., the company pled guilty to a Clean Water Act violation relating to two pipeline leaks of crude oil, one of which was the largest spill to ever occur on the North Slope. The company paid a $12 million criminal fine, $4 million in community service payments to the National Fish and Wildlife Foundation, and $4 million in criminal restitution to the State of Alaska, and will serve three years probation.

The prosecution of national corporations deters widespread violations, and encourages sector-wide compliance. In the largest criminal fine ever for a misdemeanor violation of the Clean Water Act, CITGO was sentenced to pay a $13 million fine for the negligent discharge of pollutants into two rivers in Louisiana. CITGO failed to maintain stormwater tanks and adequate stormwater storage capacity at its petroleum refinery in Sulphur, Louisiana. As a result of these failures, approximately 53,000 barrels of oil were discharged into the Indian Marais and Calcasieu Rivers following a heavy rainstorm.

LARGE FINES FOR DUMPING IN GULF OF MEXICO

Rowan Companies, Inc., a major oil and gas drilling company, pled guilty and paid a $7 million dollar fine for three Clean Water Act felonies for discharging pollutants into the Gulf of Mexico from one of its oil rigs and for failing to notify the government of the discharges. Rowan also paid $1 million for preservation and protection projects off the coasts of Texas and Louisiana. Nine supervisory employees of Rowan also pled guilty and were fined.

FY 2008 Criminal Enforcement Program Results

Cases Initiated	319
Defendants Charged	176
Sentences (years)	57
Fines and Restitution	$64 million
Judicially Mandated Projects (cost in dollars)	$12 million
Pollutant Reductions (lbs)	1.6 million

FY 2008 data source: Integrated Compliance Information System (ICIS), October 2008.

PRISON SENTENCE FOR ILLEGAL ASBESTOS REMOVAL

Cleve Allen George, the owner of the Virgin Islands Asbestos Removal Company, received 33 months in prison for multiple Clean Air Act convictions for illegal removal of asbestos-containing material at a low-income housing project and making false statements to federal agencies about air quality monitoring at the site. The owner was also sentenced

> to three years of supervised release and required to pay for baseline X-rays for exposed workers.

The sentences for those who repeat environmental crimes are often stiffer. Ronald Jagielo, owner of MRS Plating, Lockport, New York, was sentenced to 21 months incarceration and was ordered to pay $1 million in restitution and serve three years of supervised release after pleading guilty to a felony violation for disposal of hazardous wastes without a permit. This was the second felony conviction for Jagielo, who served a year in prison in 2000 after pleading guilty to illegally discharging wastes into the Lockport water treatment system where he had installed a device that hid the discharges from inspectors.

CRIMINAL ENFORCEMENT REACHES ACROSS INTERNATIONAL BOUNDARIES

Some enforcement actions involved international defendants. For instance, the National Navigation Company (NNC), an Egyptian company with offices and head-quarters in Cairo, Egypt, operated a fleet of ocean-going vessels that transports cargo, goods and people. From 2001 through 2007, engineering crews aboard vessels operated by the NNC regularly discharged oily sludge directly into oceans throughout the world.

During the investigation, EPA and the Coast Guard discovered six NNC vessels dumped thousands of gallons of waste oil and sludge in oceans around the world and falsified records to cover it up. Engineering crews routinely discharged oily sludge by installing a bypass pipe which allowed crews to pump oily sludge directly from waste tanks aboard vessels into the ocean. NNC pled guilty and paid a $4.7 million criminal fine and $2.55 million in projects for 15 felony violations of the Act to Prevent Pollution from Ships and making false statements to federal officials.

POLLUTERS PAY FOR CLEANUP: SUPERFUND ENFORCEMENT

EPA's Office of Site Remediation Enforcement manages the enforcement of EPA's national hazardous waste cleanup programs. This includes Superfund cleanups under the Comprehensive Environmental Response, Compensation, and Liability Act and cleanups at facilities that treat, store, or dispose of hazardous waste under the Resource Conservation and Recovery Act.

Superfund enforcement and other remediation agreements resulted in an estimated 100 million cubic yards of contaminated soil cleaned up and in the remediation of approximately 255 billion cubic yards of contaminated ground water.

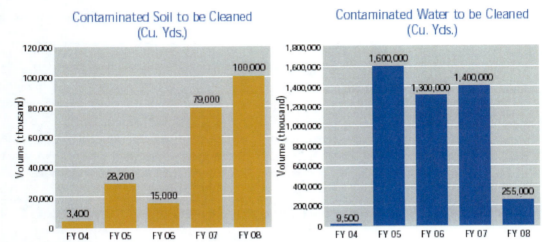

FY 2008 Data Source: Integrated Compliance Information System (ICIS), October 11, 2008; data source for previous fiscal years: ICIS. Disclaimer: Minor corrections may have been made to previous years' data.

Estimated Volume of Contaminated Soil and Water to be Cleaned Up

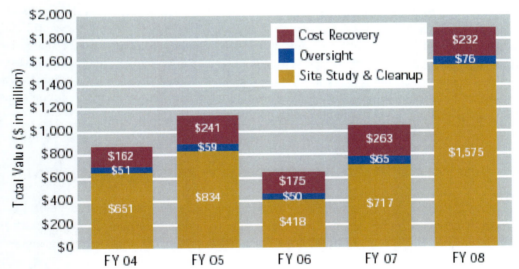

Note: All prior FY dollar figures in this chapter are adjusted to reflect the current value in FY 2008 dollars based on the monthly rate of inflation as determined by the U.S. Department of Labor Consumer Price Index for All Urban Consumers. FY 2008 Data Source for Clean-up and Cost Recovery: Comprehensive Environmental Response, Compensation & Liability Information System (CERCLIS), October 30, 2008; FY 2008 Data Source for Oversight: Integrated Financial Management System (IFMS), October 18 2008; Data source for previous fiscal years: CERCLIS and IFMS.

Private Party Commitments for Superfund Site Study & Cleanup, Oversight & Cost Recovery (Inflation Adjusted to FY 2008 Dollars)

OECA FY 2008 Accomplishments Report   155

> "Bankruptcy is not a safe haven to avoid environmental responsibilities."
> *Marcia E. Mulkey, Director Office of Site Remediation Enforcement OECA*

As a result of the Agency's efforts in FY 2008 to maximize liable party participation in performing and paying for cleanups, private parties agreed to invest approximately $1.6 billion to clean up contamination and to reimburse EPA $308 million for its past response and oversight costs.

BANKRUPT POLLUTER PAYS

EPA vigorously pursues all liable parties for Superfund cleanup costs, including bankrupt parties. In FY 2008, W. R. Grace paid $250 million to clean up asbestos contamination at the Libby Montana Superfund site. The Libby settlement sets a new record for the amount of money paid in bankruptcy to clean up a Superfund site. In addition, W. R. Grace agreed to an allowed claim in bankruptcy of $34 million for the cleanup of 32 Superfund sites in eighteen states.

ENFORCEMENT AT FEDERAL FACILITIES

EPA's actions against federal facilities secured commitments to perform cleanup work, pay penalties, and take steps to improve compliance. These actions will prevent more than 1.7 million pounds of pollutants from being released into the environment.

Cleanups at federal facilities will address more than 110 million cubic yards of contaminated soil and ground water. EPA assessed over $1.4 million in penalties and federal facilities committed to spend more than $23 million to improve facilities and operations to remedy past violations and prevent future violations.

Underground Storage Tanks

In FY 2008, EPA took nearly three dozen formal enforcement actions against federal facilities for underground storage tank (UST) violations. EPA also collected over $400,000 in penalties. Common violations included the failure to have tank release detection and tank piping.

- EPA issued a complaint to the Puerto Rico National Guard and the Army and Air Force Exchange Service, located at Camp Santiago, Puerto Rico, that proposed a penalty of $209,264 for alleged multiple violations of UST requirements.
- The U.S. Air Force, New Jersey National Guard and the Army and Air Force Exchange Service entered into a settlement with EPA resolving violations of UST requirements at the McGuire Air Force Base in New Jersey. The agreement required payment of $115,000 in penalties and the installation of proper corrosion protection equipment, overfill protection and leak detection equipment. It also requires

156 United States Environmental Protection Agency

improved annual testing and record- keeping at the 20 UST systems used to store fuels for vehicles at the base.

- The U.S. Postal Service's vehicle maintenance facility in Capital Heights, Maryland failed to install equipment that would prevent spilling and overfilling when material was transferred to the UST system. The U.S. Postal Service paid a $16,624 penalty.

Federal Superfund Sites

More than 150 federal facility cleanup sites are listed on the Superfund National Priorities List. The Superfund law requires EPA and federal owners or operators of Superfund sites to enter into enforceable agreements for the cleanup of the sites. EPA has agreements in place at most sites and continues to negotiate agreements at remaining sites.

- In FY 2008, EPA, the U.S. Navy, the U.S. Department of the Interior and the Commonwealth of Puerto Rico finalized an agreement for a former military site on the Puerto Rican island of Vieques.
- EPA and the U.S. Coast Guard also completed an agreement governing the cleanup of the Coast Guard's Curtis Bay facility in Baltimore, Maryland.
- EPA takes action when federal facilities are not complying with cleanup agreements. In FY 2008, EPA enforced against the Department of Energy (DOE) for failure to perform cleanup work at the Hanford site in Washington. DOE agreed to pay a $285,000 penalty, purchase two emergency response boats (estimated cost $200,000) for the local sheriff's office to respond to any hazardous material spills, and construct a greenhouse and nursery (estimated cost over $600,000) at a nearby campus of Washington State University to grow native vegetation to be used to rehabilitate habitat at the site. DOE also agreed to pay a $75,000 penalty for missing cleanup deadlines.
- When the U.S. Navy failed to properly monitor wells at the Brunswick Naval Air Station in Maine, EPA assessed $153,000 in stipulated penalties.
- Tyndall Air Force Base in Florida is a Superfund site where EPA found an imminent and substantial endangerment due to contamination in ground and surface water and in soil and sediments at the base. The ground water is used for drinking and nearby Shoal Bayou is used for recreational fishing and wading and has sensitive ecological resources. Because of this endangerment, EPA issued a RCRA order requiring the Air Force to investigate contamination at the base and take action to clean it up.
- Under agreements completed prior to FY 2008, federal facilities continue to investigate and clean up environmental contamination. DOE is currently cleaning up contaminated ground water and soil at an estimated cost of over $626 million at the Lawrence Livermore National Laboratory Site 300 in California, a high-explosives test facility.
- The U.S. Army will spend over $150 million to clean up almost 45 million cubic yards of contaminated ground water at Fort Ord, a former base near Monterey Bay in California. The Army will also dedicate part of the base as a wildlife reserve after munitions in the soil are addressed.

CRIMINAL ENFORCEMENT FOR ILLEGAL WASTE DISCHARGE

A former Chief Warrant Officer in the U.S. Coast Guard was sentenced in U.S. District Court in Hawaii for making a false statement to federal criminal agents investigating allegations of potential discharges of oil-contaminated waste from his Coast Guard cutter. The officer was sentenced to pay a $5,000 fine, serve 200 hours of community service and serve two years of probation. In the indictment, the officer was cited for lying to federal criminal investigators about his knowledge of an illegal discharge of bilge wastes through the ship's deep sink into Honolulu Harbor.

ENVIRONMENTAL JUSTICE FOR ALL

The Environmental Justice (EJ) program continues to assist the Agency in integrating environmental justice into key agency actions, strategic plans, and guidance. EPA's efforts support Presidential Executive Order 12898, *Federal Actions to Address Environmental Justice in Minority Populations and Low-Income Populations.*

In FY 2008, EPA continued to assist environmental justice communities on proactive, strategic, and visionary approaches to address their environmental issues. EPA collaborated with community-based organizations to achieve the following results:

- Implemented a community-led campaign in Bushwick, New York to reduce indoor exposure to asthma triggers.
- Developed a beach closure management plan for the State of Washington, and raised the community's awareness about safe and sustainable methods of harvesting shellfish.
- Cleaned up and prepared an abandoned lot for redevelopment in the Hawaiian island of Kauai.

"At EPA, environmental justice is a program, not just a slogan. . . . No other federal agency has attempted to incorporate environmental justice into its programs, policies, and activities as comprehensively as EPA."

Charles Lee, Director Office of Environmental Justice OECA

EPA'S ENVIRONMENTAL JUSTICE PRIORITIES

- Reduce asthma attacks
- Reduce exposure to air toxics
- Reduce incidences of elevated blood lead levels
- Ensure that companies meet environmental laws
- Ensure that fish and shellfish are safe to eat
- Ensure that water is safe to drink

158   United States Environmental Protection Agency

- Revitalize brownfields and contaminated sites
- Foster collaborative problem-solving

In the last three years, EPA has made significant progress in strengthening its environmental justice program through the integration of environmental justice considerations into EPA's core planning and budgeting processes. The Agency's eight national environmental justice priorities are reflected in the Agency's Strategic Plan and in FY 2008 were a focus in the annual National Program Manager Guidance documents. By instituting these actions, EPA is building a stronger foundation to successfully integrate environmental justice into its programs for the long term.

Environmental Justice Achievement Awards

In FY 2008, EPA presented its first annual awards to recognize organizations for distinguished accomplishments in addressing environmental justice issues. Projects included empowering residents to clean up New Orleans East for a safe return after Hurricane Katrina and developing a tool to target high-risk homes with lead contamination in Durham, North Carolina. EPA received dozens of nominations from across the United States. This year's twelve award recipients, listed below, include community-based organizations, universities, and state and local governments from nine states.

EPA's Inaugural Environmental Justice Awardees

- Anahola Homesteaders Council (Kauai, Hawaii)
- Center for Environmental and Economic Justice (Biloxi, Mississippi)
- Citizens for Environmental Justice (Savannah, Georgia)
- Communities for a Better Environment (Huntington Park, California)
- Dillard University, Deep South Center for Environmental Justice (New Orleans, Louisiana)
- Duke University, Children's Environmental Health Initiative (Durham, North Carolina)
- Medical University of South Carolina (Charleston, South Carolina)
- Negocio Verde Environmental Justice Task Force (County of San Diego, California)
- New Mexico Environment Department (Santa Fe, New Mexico)
- Safer Pest Control Project (Chicago, Illinois)
- South Carolina Department of Health and Environmental Control (Columbia, South Carolina)
- West End Revitalization Association (Mebane, North Carolina)

EJ Grants Program

In FY 2008, EPA announced the "Environmental Justice Small Grants Program Application Guidance FY 2008." These grants are designed for projects that address local environmental and public health issues within an EJ community. In addition, the grants aim to assist recipients in building collaborative partnerships. EPA will award the 40 small grants totaling $800,000 early next year. For more information visit our Web site at www.epa.gov/compliance/environmentaljustice/grants/ej- smgrants.html.

ENSURING COMPLIANCE IN INDIAN COUNTRY

Working with federally-recognized Indian tribes, EPA uses compliance assistance, inspections, and enforcement to address significant human health and environmental problems in Indian Country.

As part of our Indian Country priority, OECA continued to focus attention on drinking water and on solid waste issues in Indian Country in FY 2008. EPA took six enforcement actions to protect the safety of drinking water in Indian Country. These actions represent the largest number of formal enforcement actions taken in Indian Country in one year under the Safe Drinking Water Act. Examples of EPA's enforcement actions are described below.

- EPA Region 8 enforced against the Fort Belknap Community Council and Prairie Mountain Utilities for violations found at three public water systems. Another action was taken against a Northern Cheyenne Tribe for faulty water storage tanks that could potentially contaminate drinking water.
- EPA enforced against a facility operating on Arizona tribal lands in response to an imminent and substantial endangerment created by ground water contamination. EPA Region 9 issued a unilateral order to the Plymouth Tube company on the Gila River Indian Community in Chandler after the tribe discovered a contamination plume below the facility. The plume contained trichloroethylene and other solvents in concentrations above federal drinking water standards. Ground water is the sole source of drinking water for the Tribe and the order requires the company to investigate the nature and extent of the ground water.

"EPA has a responsibility to help tribes protect their resources and provide basic services like water and sewer to their members."

Wayne Nastri, Regional Administrator EPA Region 9

ENVIRONMENTAL REVIEWS MAKE A DIFFERENCE: EPA'S NEPA PROGRAM

OECA's Office of Federal Activities and its regional counterparts review and comment on other federal agencies' Environmental Impact Statements (EISs). Agencies prepare the

EISs under the National Environmental Policy Act (NEPA) and EPA reviews the documents in accordance with Section 309 of the Clean Air Act. EPA's review is intended to help federal agencies identify and ultimately avoid or mitigate potential adverse impacts from their projects.

In FY 2008, EPA reviewed over 500 EISs involving a wide range of federal projects. Some of the project reviews included: the establishment of offshore liquid natural gas ports, alternative energy projects (e.g., wind turbines), major highway projects (e.g., I-69 the NAFTA Highway), the Red River valley water supply project, and oil and gas development projects.

> "NEPA remains a valuable tool for understanding and mitigating the environmental impacts of federal actions."
>
> *Susan Bromm, Director office Federal Activities OECA*

Over 75 percent of the significant adverse effects identified through EPA's reviews of other agencies' EISs were reduced through project modifications and/or mitigation commitments. As one example, EPA's direct involvement in the proposed expansion of the Pinedale Anticline natural gas well field in Wyoming led to commitments to reduce greenhouse gas emissions and remediate ground water contamination.

EPA was successful in its first use of a third-party mediator in an EIS review of a joint U.S. Forest Service and U.S. Army Corps of Engineers proposal to expand a reservoir in Colorado. EPA Region 8 raised concerns over the potential impacts to rare and valuable wetlands in the area. The mediator provided assistance in bringing together the agencies, identifying their interests, and developing options. Ultimately the agencies agreed to include new alternatives that would protect the wetlands.

PROJECT REVIEWS IDENTIFY ENVIRONMENTAL JUSTICE IMPACTS

EPA's review of EISs on federal projects can help identify and mitigate the environmental justice concerns associated with major federal projects. For example, the Port of Los Angeles signed a resolution to adopt and implement clean air initiatives and develop a mitigation fund for projects intended to mitigate air quality impacts to the neighboring San Pedro and Wilmington communities. The resolution prevented litigation between the environmental community and the Port over the TraPac Terminal Expansion Project's air quality and health impacts to the neighboring environmental justice community.

INTERNATIONAL COMPLIANCE ACTIVITIES

EPA promotes international compliance with environmental regulations in two distinct ways. First, EPA works with state, federal, and international governments to secure compliance along the border and to ensure that imported goods and hazardous waste shipments comply with U.S. environmental laws. Second, EPA networks with other countires to share information and techniques for compliance assurance, and provides technical assistance and training to increase enforcement and compliance capacity.

> "Through our continued dialogue and targeted initiatives, EPA and our international neighbors are writing the next chapter in our countries' ongoing book of environmental collaborations."
>
> *Stephen L. Johnson, EPA Administrator*

In FY 2008, EPA reviewed and processed 1,185 hazardous waste notices for 12,184 waste streams—a new record—for imports and exports of hazardous waste. EPA's objections to certain notices prevented the environmentally unsound importation of 97 thousand metric tons of hazardous waste.

Enforcement Capacity Building

In FY 2008, international capacity building efforts marked progress with the Eighth International Conference on Environmental Compliance and Enforcement held in Cape Town South Africa and co-chaired by OECA's Principal Deputy Assistant Administrator. The conference resulted in the creation of the International Network of Environmental and Compliance Training Professionals, which will focus on international sharing of information and techniques for training environmental professionals (e.g., inspectors). The conference also launched the Seaport Environmental Security Network. This project will strengthen the enforcement capacity of both developed and developing countries to prevent illegal, hazardous waste shipments through ports, and to prevent dumping in the developing world.

Foreign Manufacturers and U.S. Importers

In FY 2008, EPA enforcement actions addressed many problems with imported products that did not comply with environmental requirements. Imports included non-road engines and parts that do not meet U.S. air pollution requirements, products containing unregistered pesticides that are harmful to children and chemicals that deplete the stratospheric ozone layer (e.g., confetti string, consumer products manufactured with radioactive scrap metals, lead in faucets).

EPA is working directly with the Chinese government on import safety issues. In December 2007, EPA and China's General Administration of Quality Supervision, Inspection and Quarantine signed a Memorandum of Understanding which provides a framework for cooperation to protect human health and the environment in the field of imported and exported products.

ILLEGAL MANUFACTURE AND IMPORTATION

In FY 2008, EPA reached a landmark settlement with a Taiwanese manufacturer and three U.S. corporations (MTD) to resolve violations arising from the illegal manufacture and importation of approximately 200,000 chainsaws that failed to meet federal air pollution standards. The foreign manufacturers and U.S. importers of these chainsaws agreed to pay a $2 million civil penalty. The defendants also agreed to spend approximately $5 million on projects to reduce air pollution.

TIPS AND COMPLAINTS

EPA's tips and complaints Web site (www.epa.gov/tips) is an important tool for identifying potential environmental violations. Established in January 2006, our easy-to-spot icon enables concerned citizens and employees to report potential violations in their communities or workplaces.

In FY 2008, EPA received a total of 7,835 tips. Tips are reviewed by EPA's enforcement programs to determine potential civil or criminal violations. Since the launch of the Web site, 1,300 potentially criminal tips have been referred to field offices and 19 tips resulted in criminal cases.

Two cases resulted in convictions during FY 2008. The City of Lake Ozark, Missouri, paid a $50,000 fine after pleading guilty to discharging a pollutant without a permit into the Lake of the Ozarks. Its Public Works director also pled guilty to one count of failing to report a sewage discharge.

In the second case, an official of Environmental Staffing Acquisition Corporation, a company that provided temporary workers for environmental cleanup projects will serve two years on federal probation for creating documents that falsely purported to certify that the company's employees had received medical evaluations required by the U.S. Occupational Health and Safety Administration.

Appendix A: Organizational Chart

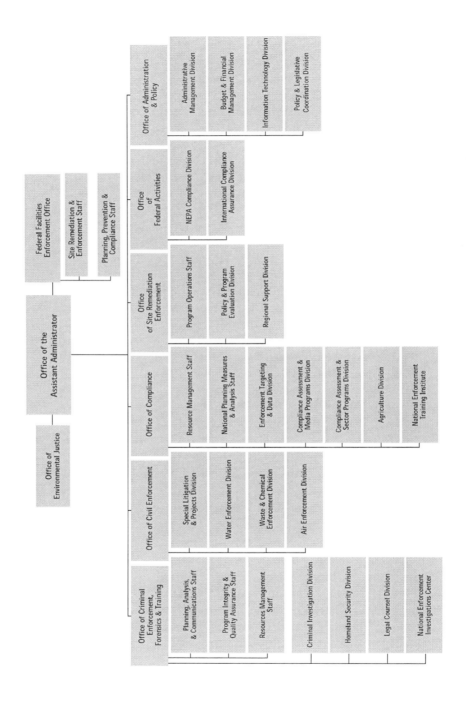

Office of Enforcement and Compliance Assurance (OECA)

Appendix B: Numbers at a Glance

FY 2008

Results Obtained From EPA Civil Enforcement Actions:
- Estimated Direct Environmental Benefits

Direct Environmental Benefits

Pollution Reduced, Treated or Eliminated (Pounds)[1]	3,900,000,000
Hazardous Waste Reduced, Treated, or Properly Disposed of (Pounds)[1,2]	6,500,000,000
Contaminated Soil to be Cleaned Up (Cubic Yards)	100,000,000
Contaminated Water to be Cleaned Up (Cubic Yards)	255,000,000
Stream Miles Protected (Linear Feet)	53,000
Wetlands Protected (Acres)	5,200
People Protected by Safe Drinking Water Act Enforcement (# of People)	1,024,000
Thermal Pollution Reduced (Water) (MMBTUs)[3]	40,300,000

Preventative Environmental Benefits

Hazardous Waste Prevented from Release (Cubic Yards)	220,000
Underground Storage Tank Capacity Prevented from Release (Gallons)	1,300,000
People Notified of Potential Drinking Water Problems (# of People) Underground Injection Wells Prevented from Leaking (# of Wells)	115
PCB Disposal Corrected (Cubic Yards)	900
Lead-Based Paint Contamination Prevented (# of Housing Units, Schools, Buildings)	15,000
Volume of Oil Spills Prevented (Gallons)	194,000,000
Pesticides or Pesticide Products Prevented from Distribution, sale or Use due to Mislabeling or Improper Registration (Pounds)	50,000,000

- Investments in Pollution Control and Clean-up (Injunctive Relief)
- $11,700,000,000
- Investments in Environmentally Beneficial Projects (SEPs) $39,000,000
- Civil Penalties Assessed

Administrative Penalties Assessed	$38,200,000
Judicial Penalties Assessed	$88,400,000
Stipulated Penalties Assessed	

- EPA Civil Enforcement and Compliance Activities

Referrals of Civil Judicial Enforcement Cases to Department of Justice (DOJ)	280
Supplemental Referrals of Civil Judicial Enforcement Cases to DOJ	35
Civil Judicial Complaints Filed with Court	164
Civil Judicial Enforcement Case Conclusions	192
Administrative Penalty Order Complaints	2,056
Final Administrative Penalty Orders	2,084
Administrative Compliance Orders	1,390
Cases with SEPs	188

- EPA Compliance Monitoring Activities

Inspections/Evaluations	20,000

Civil Investigations	222
Number of Regulated Entities Taking Complying Actions during EPA Inspections/Evaluations	1,100
Number of Regulated Entities Receiving Compliance Assistance during EPA Inspections/Evaluations	11,600
Inspections Conducted by Tribal Inspectors Using Federal Credentials[4]	334

- **EPA Superfund Cleanup Enforcement**

Percent of non-Federal Superfund Sites with Viable, Liable Parties where an Enforcement Action was Taken Prior to the Start of the Remedial Action	100%
Private Party Commitments for Site Study and Cleanup (including cash outs)	$1,575,000,000
Private Party Commitments for Oversight	$76,000,000
Private Party Commitments for Cost Recovery	$232,000,000
Percent of Cost Recovery Cases Greater Than or Equal to $200,000 that were Addressed before the Statute of Limitations Expired	100%

- **EPA Criminal Enforcement Program**

Years of Incarceration	57
Fines and Restitution	$63,500,000
Value of Court Ordered Environmental Projects	$12,000,000
Environmental Crime Cases Initiated	319
Defendants Charged	176
Estimated Pollution Reduced, Treated or Eliminated Commitments (Pounds)[1]	1,600,000

- **EPA Voluntary Disclosure Program**

Estimated Pollution Reduction Commitments Obtained as a Result of Voluntary Disclosures (Pounds)	5,400,000
Voluntary Disclosures Initiated (Facilities)	2,294
Voluntary Disclosures Resolved (Facilities)	640
Voluntary Disclosures Initiated (Companies)	538
Voluntary Disclosures Resolved (Companies)	451
Notices of Determination (NODs)	364

- **EPA Compliance Assistance**

Total Entities Reached by Compliance Assistance	361,000
Number of User Visits to Web-Based Compliance Assistance Centers	2,220,000

The primary source for the data displayed in this document is the Regions' certified FY 2008 end of year workbooks as of November 5, 2008. The official databases of record are: Integrated Compliance Information System (ICIS), Criminal Case Reporting System, Comprehensive Environmental Response, Compensation & Liability Information System (CERCLIS), Resource Conservation and Recovery Act Information (RCRAInfo), Air Facility System (AFS), and Permit Compliance System (PCS).

APPENDIX C: ABBREVIATIONS & ACRONYMS

CAA	Clean Air Act
CAFOs	Concentrated Animal Feeding Operations
CERCLA	Comprehensive Environmental Response, Compensation and Liability Act (aka "Superfund")
CWA	Clean Water Act
EJ	Environmental Justice
EPA	Environmental Protection Agency
EPCRA	Emergency Planning & Community Right-to-Know Act
FIFRA	Federal Insecticide, Fungicide and Rodenticide Act
M PRSA	Marine Protection, Research, and Sanctuaries Act
NSR	New Source Review
OECA	Office of Enforcement and Compliance Assurance
PSD	Prevention of Significant Deterioration
RCRA	Resource Conservation & Recovery Act

End Note

* Note: All prior FY dollar figures in this chapter are adjusted to reflect the current value in FY 2008 dollars based on the monthly rate of inflation as determined by the U.S. Department of Consumer Price Index for All Urban Consumers.

[1] Projected reductions to be achieved during the one year period after all actions required to attain full compliance have been completed.

[2] In FY 2008, for the first time, OECA is piloting a new Environmental Benefits outcome reporting category to count pounds of "Hazardous Waste Treated, Minimized or Properly Disposed Of " from enforcement cases. OECA has determined that none of the previously established outcome categories are appropriate for counting the environmental benefits obtained from EPA's hazardous waste cases. For FY 2008, this new pilot category includes only results from RCRA cases, but, in the future, similar results obtained from enforcement actions under other statutes, particularly CERCLA, may also be included.

[3] In FY 2008, for the first time, OECA is including a new Environmental Benefit outcome reporting category to count British Thermal Units (BTUs) of "Thermal Pollution Reduced (Water)". OECA has determined that none of the previously established outcome categories is appropriate for counting the environmental benefits obtained from enforcement cases that produce reductions in thermal pollution. An MM BTU equals one million (1,000,000) BTUs.

[4] In FY 2008, for the first time, OECA is creating a separate reporting category to count the number of tribal inspections conducted by tribal inspections using federal credentials. Inspections conducted by tribal inspectors using federal credentials are done "on behalf" of the Agency, but are not an EPA activity.

INDEX

A

abatement, 17
accountability, 40, 49, 70, 96
accounting, 36
accuracy, 34, 38, 40, 42, 43
acid, 9, 145, 148
acidic, 149
Act to Prevent Pollution from Ships, 15, 153
acute, 148
adjudications, 28
administration, 59, 69, 135
administrative, viii, 10, 11, 28, 30, 34, 36, 42, 49, 70,
    97, 101, 102, 103, 111, 112, 113, 114, 117, 118,
    119, 126, 127, 128, 135, 141
Administrative Procedure Act, 13, 133, 134
administrators, 73
adults, 28
AEP, 2, 5
age, 120
agent, 27
agents, 108, 115, 116, 151, 157
aging, 2, 5, 27, 30
agricultural, 12, 83
agriculture, 108, 113, 121
aid, 10, 50, 67, 89, 155
air, vii, viii, 1, 2, 3, 5, 7, 17, 20, 21, 40, 53, 54, 63,
    65, 67, 68, 72, 84, 87, 89, 90, 97, 98, 102, 103,
    109, 112, 113, 126, 127, 134, 137, 140, 142, 145,
    146, 147, 148, 152, 157, 160, 161
air emissions, 68, 98, 109
Air Force, 27, 155, 156
air pollutants, 3, 5, 20, 40, 65, 109, 146
air pollution, 2, 5, 63, 72, 134, 148, 161
air quality, 21, 63, 65, 84, 152, 160
air toxics, 65, 157
aircraft, 10

Alabama, 27, 29
Alaska, 18, 22, 81, 152
aliens, 26
alternative, 22, 77, 91, 93, 160
alternative energy, 160
alternatives, 160
ambient air, 63
amendments, 24, 57, 64, 65, 66, 83, 101
American Indian, vii, 1, 4, 107
American Indians, vii, 1, 4
animal feeding operations, 68, 94
animal waste, 95, 146
APA, 13
Appalachia, 138, 149
appellate courts, 14, 26, 29
appendix, 60, 112
application, 30, 81, 95, 127
appropriations, 66, 98, 100, 122, 124, 132, 136
aquatic habitat, 12
aquifer, 14
Arizona, 7, 11, 14, 26, 27, 56, 65, 67, 69, 70, 77, 81,
    159
Arkansas, 56, 57, 64, 69, 77
Army, 7, 10, 12, 14, 16, 21, 24, 26, 27, 103, 104,
    115, 155, 156, 160
Army Corps of Engineers, 7, 12, 14, 16, 24, 26, 104,
    115, 160
asbestos, 2, 9, 17, 115, 147, 152, 155
Asia, 143
Asian, 25
assessment, 11, 22, 56, 57, 58, 71, 74, 75, 76, 78, 79,
    81, 84, 92, 105, 113, 117, 118, 135, 136
assets, 7, 29
Association of Southeast Asian Nations, 25
asthma, 5, 147, 148, 157
asthma attacks, 157
Atlantic, 8, 22, 29
atmosphere, 41
attacks, 148, 157

168 Index

Attorney General, 4, 20, 23, 41, 49, 98, 151
auditing, 35, 57, 78, 101, 150
Austria, 25
authority, viii, 16, 25, 28, 34, 54, 59, 60, 64, 72, 73, 83, 85, 87, 88, 96, 103, 105, 106, 107, 111, 112, 117, 118, 119, 125, 133, 134
autonomy, ix, 55, 89, 91, 97, 99
availability, 102, 128
avoidance, 117, 118
awareness, 112, 157

B

bacteria, 146
Bahrain, 25
bankruptcy, 9, 155
barriers, 81
basic services, 159
benefits, 4, 12, 23, 29, 30, 38, 58, 70, 81, 84, 94, 101, 111, 119, 121, 125, 138, 139, 147, 148, 166
benzene, 20
Best Practice, 31, 133
bilge, 157
binding, 50
BLM, 13
blood, 157
blood lead levels, 157
boats, 156
bottom-up, 92, 95
British Petroleum, 18, 152
British Thermal Units (BTUs), 166
bronchitis, 148
BTUs, 166
budgetary resources, 72
Bureau of Land Management, 13
bypass, 153

C

cadmium, 16
CAFO, 149
cancer, 20
capacity building, 25, 161
carbon, 63, 134
carbon monoxide, 63, 134
carcinogen, 8
cargo, 3, 15, 153
case law, 25, 118
catfish, 19
CBS, 10
cement, 145
channels, 12

chemical weapons, 27, 134
chemicals, 18, 59, 115, 127, 147, 161
childhood, 147, 148
children, 161
China, 19, 25, 138, 161
chloride, 8
chromium, 16
citizens, 4, 22, 98, 103, 108, 133, 162
civil action, 68
Civil Rights Act, 135
clean air, 65, 160
Clean Air Act, vii, 2, 3, 5, 33, 38, 49, 53, 63, 64, 65, 68, 72, 77, 81, 83, 84, 87, 88, 90, 98, 102, 105, 128, 129, 131, 133, 134, 135, 145, 146, 147, 152, 159, 166
Clean Water Act, vii, 2, 3, 6, 33, 42, 53, 57, 63, 64, 68, 77, 81, 83, 84, 87, 88, 92, 96, 98, 102, 105, 128, 129, 131, 133, 134, 135, 146, 152, 166
cleaning, 84, 156
cleanup, viii, ix, 2, 8, 9, 10, 21, 66, 84, 97, 105, 118, 134, 135, 137, 138, 141, 145, 147, 153, 155, 156, 162
climate change, 22
closure, 11, 117, 147, 157
Co, 2, 7, 8, 9, 10, 13, 15, 17, 25, 132
coal, 2, 5, 7, 22, 40, 42, 138, 145, 148, 149
coal mine, 149
Coast Guard, 19, 27, 104, 115, 153, 156, 157
collaboration, 70
collateral, 10
colleges, 138, 143
Colorado, 9, 19, 23, 31, 74, 79, 81, 132, 160
Columbia River, 4, 11, 28
combined effect, 22
commerce, 12, 98
Committee on Appropriations, viii, 53, 85, 86
Committee on Environment and Public Works, 53, 132
communication, 70, 136
communities, 52, 57, 67, 69, 76, 83, 142, 157, 160, 162
community, ix, 15, 16, 17, 18, 59, 72, 74, 75, 88, 93, 97, 101, 103, 105, 108, 109, 110, 112, 120, 124, 133, 136, 139, 142, 152, 157, 158, 160
community service, 15, 16, 17, 18, 152, 157
compensation, 14
competition, 81
competitive advantage, 49, 72, 117
compilation, 112
complexity, 100
components, 72, 80

Comprehensive Environmental Response, Compensation, and Liability Act (CERCLA), 9, 10, 27, 98, 105, 109, 110, 117, 126, 131, 153, 166
concealment, 18
concentrates, 92
concrete, 31, 148
conditioning, 54
confidence, 81
confinement, 17
Congress, 4, 20, 21, 22, 23, 24, 28, 34, 35, 40, 42, 43, 49, 50, 59, 64, 65, 75, 95, 96, 97, 98, 99, 100, 101, 102, 103, 107, 110, 115, 116, 119, 122, 124, 133, 136, 143
Congressional Budget Office, 77
congressional hearings, 101, 119
Congressional Record, 132, 136
Connecticut, 84, 138
connectivity, 30
consent, 2, 3, 5, 6, 7, 8, 9, 11, 25, 43, 50, 106, 117
conservation, 4, 7, 11, 14, 31, 149
Consolidated Appropriations Act, 132, 136
consolidation, 126
conspiracy, 16, 17, 19, 116, 151
constraints, 67, 95
construction, 3, 6, 26, 57, 64, 65, 105, 108, 146, 148
construction sites, 6, 146, 148
consultants, 121
Consumer Price Index, 149, 150, 154, 166
contaminants, 6, 21, 27, 146
contamination, 2, 9, 52, 57, 67, 69, 84, 98, 110, 134, 155, 156, 158, 159, 160
conversion, 30, 50
conviction, 115, 135, 153
copper, 6
corporations, 5, 108, 152, 161
corridors, 22
corrosion, 18, 155
corrosive, 17
cost accounting, 92
cost saving, 122
cost-effective, 94
costs, 2, 8, 9, 10, 11, 20, 38, 40, 41, 49, 50, 53, 63, 65, 72, 117, 118, 134, 138, 141, 150, 155
costs of compliance, 41
cotton, 151
Council on Environmental Quality, 24
counsel, 30, 98, 118
Court of Appeals, 50
courts, vii, 1, 3, 6, 36, 98, 111, 117, 138
covering, 74
credentials, 133, 145, 166
credit, 59, 95
crime, 115

crimes, 16, 115, 116, 151, 153
critical habitat, 11
criticism, 74, 79, 107, 120
crude oil, 18, 152
cryptosporidium, 21
customers, 16, 19

D

danger, 111
data collection, 64, 78
database, 35, 109, 120, 125, 127, 144
death, 149
decision makers, 75
decision-making process, 69
decisions, vii, 1, 12, 29, 41, 49, 55, 92, 94, 96, 111, 114, 120, 140
defendants, 1, 5, 7, 9, 10, 15, 16, 19, 38, 41, 43, 138, 150, 153, 162
defense, 14
defenses, 13
definition, 42
deflate, 83
delivery, 22
demographic characteristics, 120
demographics, 126
denial, 20, 30, 117
density, 13
Department of Defense (DOD), 101, 134
Department of Energy (DOE), 10, 101, 156
Department of Homeland Security, 3, 26, 104
Department of Justice (DOJ), vii, 1, 4, 33, 41, 98, 100, 104, 136, 138, 140, 151, 164
Department of State, 25
Department of the Interior, 23, 29, 103, 156
Department of Transportation, 23, 104
destruction, 13, 27
detection, 84, 155
detention, 17, 18
deterrence, 91, 101, 151
developing countries, 161
diesel, 38
dioxins, 147
directives, 69, 134
discharges, 3, 6, 15, 16, 98, 102, 109, 146, 152, 153, 157
disclosure, 80, 121, 150
discriminatory, 136
diseases, 148
disinfection, 21
disputes, 22, 29
dissenting opinion, 16
distribution, 77, 92, 95, 120, 122

170   Index

District of Columbia, 6, 50, 83, 84
District of Columbia Circuit, 50
diversity, 13, 108, 109, 124
division, 18, 89, 117, 133
DOT, 22, 23, 104, 115
draft, 50, 58, 60, 81
drainage, 9
drinking, 21, 28, 107, 108, 113, 115, 147, 156, 159
drinking water, 21, 28, 107, 108, 113, 115, 147, 159
drought, 4, 23
due process, 10
dumping, 107, 147, 161
DuPont de Nemours, 10
duties, 21, 29, 55, 78, 108

E

earnings, 108
ecological, 156
ecology, 23
economic activity, 29
ecosystem, 12, 147
elderly, 5
electric power, 2, 5, 89, 148
electric utilities, 63
electricity, 22
elementary school, 17
elk, 19
email, 30
Emergency Planning and Community Right-to-
  Know Act (EPCRA), 126, 129, 131, 166
emergency response, 156
emission, 65, 102, 134
emitters, 108
employees, 18, 30, 31, 60, 88, 92, 95, 108, 140, 152,
  162
Endangered Species Act, 4, 11
energy, 12, 22, 30, 31, 84
energy efficiency, 30
Energy Policy Act, 22, 66, 84
Energy Policy Act of 2005, 66, 84
engagement, 29
engines, 143, 161
enterprise, 30
environment, vii, viii, 1, 6, 18, 27, 33, 34, 41, 50, 53,
  59, 66, 68, 74, 76, 89, 97, 102, 107, 108, 110,
  113, 124, 138, 139, 144, 147, 149, 152, 155, 161
environmental audit, 121
environmental conditions, 54, 75, 76, 94, 106, 113,
  120
environmental contamination, 9, 156
environmental impact, 12, 22, 84, 120, 160
environmental issues, 68, 102, 107, 120, 157

Environmental Management Systems, 121
environmental policy, 132
environmental protection, viii, 74, 79, 85, 87, 107,
  115
environmental regulations, 74, 86, 87, 93, 104, 126,
  160
environmental standards, 94, 114
equities, 3
equity, 120
erosion, 95
estimating, 40
Everglades, 12, 14
examinations, 87, 145
Executive Order, 41, 120, 133, 135, 157
exercise, 14, 89, 91
expenditures, 111, 117, 124
expertise, 31, 69, 72, 75, 76, 81, 93, 104
explosives, 156
exports, 161
exposure, 8, 120, 157
Exxon, 8

F

failure, 14, 15, 18, 20, 69, 125, 146, 155, 156
fairness, 74, 86, 120
false statement, 15, 16, 17, 18, 116, 151, 152, 153,
  157
Federal Bureau of Investigation, 27, 104, 115
federal government, 8, 9, 12, 24, 27, 36, 38, 43, 49,
  50, 96, 99, 106, 108, 110, 117, 123, 151
Federal Highway Administration, 22
Federal Insecticide, Fungicide, and Rodenticide Act,
  (FIFRA), 77, 102, 105, 113, 126, 127, 129, 131,
  166
federal law, 30, 33, 49, 59, 60, 83, 105, 106
Federal Register, 20, 67, 78, 133, 135, 136
federalism, 99, 132
fee, 63, 65
feedback, 69, 81, 106
feeding, 95
fees, 63, 65, 70, 84, 110
felony, 153
fencing, 26
finance, 8, 9, 63
financial resources, 52, 57, 66, 67, 69, 86, 88, 147
fines, vii, viii, 1, 16, 91, 95, 97, 101, 110, 115, 119,
  130, 138, 142, 151
fire, 2, 5, 13, 27, 40, 105, 147, 148
fish, 4, 19, 23, 24, 28, 138, 149, 157
Fish and Wildlife Service (FWS), 11, 104, 115
fisheries, 24
fishing, 4, 11, 19, 28, 156

Index

171

flexibility, 4, 23, 29, 55, 58, 59, 70, 71, 75, 82, 89, 90, 91, 95, 100, 111
flood, 12, 21
flooding, 138
flow, 148
fluctuations, 112
focusing, 111, 145
forensic, 151
forest management, 13
forest restoration, 13
Forest Service, 13, 14, 160
forestry, 13, 25
forests, 4, 5, 13
France, 25
fraud, 16, 151
fresh water, 28
friction, 100
fuel, 13
funding, ix, 15, 25, 49, 52, 55, 57, 60, 61, 62, 63, 64, 66, 70, 74, 75, 76, 92, 96, 97, 98, 99, 100, 101, 106, 117, 122, 123, 124, 136, 147
funds, 29, 49, 52, 54, 57, 58, 62, 63, 66, 70, 84, 94, 96, 147
FWS, 11, 13, 14, 22, 24

G

garbage, 83
gas, 4, 22, 29, 65, 67, 152, 160
gasoline, 8, 63
GDP, 83, 130
gender, 120
generators, 66
Georgia, 8, 20, 27, 77, 158
glass, 145
goals, 34, 40, 55, 69, 88, 90, 91, 92, 93, 95
Government Accountability Office (GAO), 33, 49, 51, 85, 96, 99, 132
Government Performance and Results Act (GPRA), 40, 50, 92
grants, 52, 54, 56, 57, 58, 62, 63, 64, 65, 70, 75, 106, 120, 124, 158
gravity, 34, 72, 80, 117
grazing, 4, 29
greenhouse, 20, 156, 160
greenhouse gas, 20, 160
Gross Domestic Product (GDP), 83
ground water, 141, 146, 147, 153, 155, 156, 159, 160
groundwater, 16, 23, 28, 113
groups, 2, 5, 13, 34, 50, 74, 98, 103, 119, 120, 121, 124
growth, 13, 52, 57, 60, 83
Guatemala, 25

guidance, 34, 40, 42, 53, 54, 56, 58, 59, 69, 70, 71, 75, 77, 78, 79, 81, 90, 103, 111, 117, 119, 121, 133, 135, 138, 157
guidelines, 115, 118
guilty, 15, 16, 17, 18, 19, 152, 153, 162
Gulf of Mexico, 18, 152

H

habitat, 4, 11, 12, 13, 28, 138, 156
handling, 25, 52, 57, 60, 67, 69, 126, 147
harm, 3, 28, 34, 72, 102, 115, 117, 150, 152
harvesting, 157
Hawaii, 3, 6, 26, 157, 158
hazardous materials, 17, 115, 147
hazardous substance, 52, 57, 67, 69, 134
hazardous substances, 52, 57, 67, 69, 134
hazardous wastes, 17, 66, 98, 153
haze, 148
health, vii, 5, 8, 29, 33, 68, 74, 84, 101, 102, 113, 119, 120, 124, 138, 144, 147, 149, 151, 152, 159, 160, 161
health effects, 8
hearing, 22, 87
heart, 148
heart attack, 148
heat, 30
heavy metal, 16, 147
heavy metals, 16, 147
high-level, 21
high-risk, 158
hiring, 92, 115, 116
Homeland Security, 3, 26, 104, 116, 122, 123
House, viii, 33, 53, 85, 86, 100, 119, 122, 135
housing, 17, 152
HUD, 17, 24
human, vii, 8, 33, 68, 74, 88, 95, 102, 113, 119, 120, 124, 138, 144, 147, 149, 151, 152, 159, 161
human activity, 120
human capital, 88, 95
humans, 146
hunting, 19, 28
Hurricane Katrina, 21, 158
hydro, 6, 146
hydrocarbons, 6, 146
hydrology, 23
hydropower, 12

I

ICE, 104
Idaho, 4, 13, 28, 31

172 Index

IDEA, 127
identification, 80, 101, 126
identity, 70
Illinois, 8, 20, 27, 158
Immigration Act, 26
Immigration and Customs Enforcement, 104
immunity, 21, 118
implementation, 24, 42, 56, 57, 58, 59, 71, 79, 81, 83, 93, 100, 106, 107, 115, 120, 133
imports, 143, 161
imprisonment, vii, viii, 34, 115
incarceration, 2, 15, 16, 17, 19, 111, 117, 142, 151, 153
incentive, 80, 97, 101, 102, 103, 121, 122, 136
incentives, 93, 121, 136, 150
incineration, 27
inclusion, 119
income, 120, 152
increased workload, 58, 71
Indian, 4, 14, 28, 29, 54, 83, 107, 121, 133, 145, 147, 152, 159
Indian Gaming Regulatory Act, 29
Indiana, 7, 23
Indians, 28, 30
indication, 109
indicators, 50, 80, 86, 87, 88, 92, 101, 124
Indonesia, 25
industrial, 56, 57, 64, 68, 83, 108, 112, 146
industrial sectors, 56, 68
industrial wastes, 146
industry, 2, 3, 5, 7, 15, 20, 24, 69, 72, 83, 119, 121, 142
inequity, 120
inflation, 34, 35, 36, 38, 40, 43, 49, 50, 52, 57, 60, 63, 83, 98, 100, 130, 131, 136, 149, 150, 154, 166
information sharing, 30
information system, 94, 126, 127, 140, 143, 144, 148, 150, 152, 154, 165
infrastructure, ix, 30, 97, 99, 101
initiation, 114
injection, 21
injunction, 3, 13, 26
injury, 13, 14, 115
injustice, 120
innovation, 80, 121
insight, 124
inspection, 54, 58, 59, 64, 65, 66, 67, 71, 72, 80, 87, 89, 110, 112, 126, 161
inspections, 54, 56, 57, 64, 65, 66, 71, 74, 80, 81, 84, 86, 87, 88, 89, 90, 105, 106, 113, 127, 133, 142, 144, 145, 148, 159, 166
Inspector General, 42, 50, 55, 64, 66, 72, 73, 74, 75, 77, 83, 84, 88, 89, 92, 96, 99, 132

inspectors, 57, 64, 67, 69, 104, 112, 142, 144, 145, 153, 161, 166
instructors, 25
insurance, 10
integration, 126, 127, 158
integrity, 6, 49, 59, 96
interaction, 104
interactions, 99
interdependence, 55
interest groups, 74, 98
interference, 28, 30
internal controls, 57, 78
Internal Revenue Service, 104
international law, 15
international trade, 23
interstate, 4, 19, 84, 110
intervention, 79
interviews, 56, 77, 78
investigations, 33, 104, 116, 129, 132, 151, 165
irrigation, 12
IRS, 104, 115
island, 156, 157
ivory, 24

J

jobs, 30
judge, 26, 50, 114
judges, 25, 135
judgment, 7, 8, 9, 12, 35, 40, 50
jurisdiction, 3, 10, 13, 15, 26, 27, 42, 76, 83, 89, 91, 107, 136
jurisdictions, 86
jury, 16
justice, 16, 68, 119, 120, 139, 157, 158, 160
Justice Department, 151

K

Kentucky, 138, 149

L

Lacey Act, 19, 24
land, vii, 1, 7, 12, 24, 26, 27, 28, 29, 53, 66, 83, 87, 102, 105, 107, 126, 137, 140, 145, 149
land acquisition, 26
land use, 105
language, 29, 100, 110, 120, 124
Laos, 25
laundering, 19

Index

law, 13, 15, 23, 25, 28, 30, 31, 50, 59, 83, 87, 102, 105, 106, 114, 115, 116, 117, 118, 133, 134, 141, 143, 151, 156
law enforcement, 25, 115, 116, 151
lawsuits, 10, 13, 49, 50, 103, 114
lawyers, 104
leaks, 84, 147, 152
legislation, 15, 24, 55, 59, 78, 101, 105, 119, 124, 133, 134
legislative proposals, 24
lifetime, 6
limitations, 13, 30, 58, 71, 74, 106, 113
linear, 50
linear regression, 50
links, 136
litigation, vii, 1, 2, 13, 24, 25, 26, 28, 29, 30, 31, 41, 50, 101, 104, 106, 123, 160
liver, 8
liver cancer, 8
loan guarantees, 24
lobsters, 19
local authorities, 105
local community, 74
local government, 65, 87, 98, 103, 105, 107, 118, 121, 125, 158
Louisiana, 18, 21, 152, 158
low-income, 152
lung, 148
lung disease, 148
LUST, 123
lying, 157

measurement, 79, 120
measures, 1, 2, 3, 5, 6, 7, 11, 17, 18, 34, 43, 50, 59, 79, 80, 88, 90, 92, 94, 98, 106, 113, 140, 145, 149
media, 67, 70, 79, 80, 81, 102, 112, 125, 127
mediation, 9
melt, 146
membership, 84
mercury, 41, 52, 57, 67, 69, 142
messages, 31
metals, 16, 146, 161
metric, 161
Mexico, 3, 19, 26, 158
microbial, 21
military, 3, 4, 26, 27, 134, 156
Minerals Management Service, 22
mining, 2, 7, 9, 22, 29, 146
Minnesota, 56, 58, 64, 67, 69, 70, 77
minors, 84
missions, 65, 113, 120
Mississippi, 8, 27, 158
Missouri, 12, 81, 162
models, 118
modernization, 94, 126
money, 9, 19, 70, 84, 117, 118, 155
money laundering, 19
Montana, 2, 8, 9, 17, 65, 79, 155
motion, 7, 20, 22
MRS, 153
multimedia, 89, 112, 127
multiplicity, 124
municipal solid waste, 143

M

magnesium, 10
Magnuson-Stevens Fishery Conservation and Management Act, 11
mail fraud, 16, 151
Maine, 84, 156
maintenance, 3, 6, 12, 42, 50, 115, 146, 156
Malaysia, 25
management, 13, 24, 29, 31, 42, 55, 59, 60, 66, 67, 68, 69, 78, 80, 81, 86, 87, 88, 92, 94, 95, 106, 108, 134, 146, 149, 157
management practices, 29, 146
mandates, 69, 100
manufacturer, 5, 38, 161
manufacturing, 108, 121
maritime, 2, 15
MARPOL, 15
Maryland, 23, 81, 132, 138, 156
Massachusetts, 56, 58, 69, 70, 77, 84, 138
maximum penalty, 36

N

NAFTA, 160
nation, vii, viii, 1, 2, 3, 4, 5, 6, 7, 22, 26, 51, 59, 68, 76, 85, 87, 88, 137, 139, 142, 144, 145, 148
National Ambient Air Quality Standards, 134
National Guard, 155
National Marine Fisheries Service, 11
National Oceanic and Atmospheric Administration (NOAA), 14, 104
national origin, 136
National Park Service, 23
national parks, 4
national policy, 80, 81
national security, 3, 27, 116, 134
Native American, 133
natural, 1, 3, 4, 8, 11, 117, 123, 124, 151, 160
natural gas, 160
natural resources, vii, 1, 3, 4, 117, 123, 124, 151
Navy, 3, 26, 156
Nebraska, 81

negotiating, 105, 118
negotiation, 11, 80, 119
nerve, 27
network, 30
neurological disorder, 8
Nevada, 4, 10, 13, 23, 29
New England, 84
New Jersey, 12, 77, 81, 84, 138, 155
New Mexico, 26, 158
New Orleans, 158
New York, 12, 24, 52, 53, 56, 58, 62, 66, 69, 70, 77, 84, 132, 138, 153, 157
nitrogen, 134, 145, 147
nitrogen dioxide, 134
nitrogen oxides, 145, 147
non-biological, 21
nongovernmental, 134
non-profit, 34
North America, 12
North Carolina, 11, 132, 158
Northeast, 84
Norway, 25
nuclear, 134
Nuclear Regulatory Commission, 18
nuclear weapons, 134
nutrients, 21, 146

O

objectivity, 81
obligation, 15, 29
obligations, 24, 25, 27, 69, 87, 101, 120, 138, 142
observations, viii, 85, 86, 88, 112
obstruction, 15, 16
occupational, 120
oceans, 2, 15, 153
OCR, 135
Office of Management and Budget (OMB), 38, 77, 78, 95, 130
Office of the Chief Financial Officer, 61, 78, 95
offshore, 18, 160
Ohio, 10, 31, 50, 135
oil, 4, 8, 15, 17, 18, 22, 29, 146, 148, 152, 153, 157, 160
oil production, 18
oil spill, 8
oils, 147
Oklahoma, 81
omnibus, 132
Oregon, 4, 12, 21, 27, 28, 31, 56, 69, 70, 77
organic, 145, 146
organic compounds, 145

oversight, viii, ix, 9, 49, 51, 52, 54, 56, 58, 59, 68, 71, 73, 74, 75, 76, 77, 78, 79, 81, 84, 89, 90, 93, 95, 96, 97, 99, 100, 105, 106, 109, 124, 126, 138, 155
oxides, 147
Ozarks, 162
ozone, 134, 161

P

Pacific, 3, 11, 13, 15, 16, 83, 84
particulate matter, 134, 145, 147
partnership, 2, 25, 52, 53, 55, 56, 58, 59, 65, 67, 75, 106, 118, 121, 124
partnerships, 56, 106, 121, 122, 124, 159
pathogens, 21, 146
PCBs, 147
PCS, 127, 165
penalty, 2, 5, 6, 7, 8, 14, 15, 34, 35, 38, 41, 50, 64, 72, 73, 74, 80, 83, 98, 111, 112, 114, 117, 119, 126, 133, 138, 148, 149, 155, 156, 162
Pennsylvania, 27
periodic, 95
permit, 2, 7, 14, 17, 20, 57, 64, 65, 66, 84, 102, 109, 110, 112, 114, 126, 138, 148, 149, 153, 162
Peru, 25
pest control, 12
pest management, 108
pesticide, 105
pesticides, 105, 146, 147, 148, 161
petitioners, 20, 21, 27
petroleum, 3, 5, 8, 15, 18, 53, 68, 109, 152
petroleum products, 15
pH, 146
philosophical, 90
philosophy, 74, 86, 87
pilots, 81
pipelines, 8
planned action, 86, 88
planning, viii, 50, 51, 52, 55, 56, 57, 67, 69, 70, 76, 78, 80, 86, 88, 90, 92, 93, 94, 95, 100, 103, 106, 109, 119, 132, 136, 158
plants, 2, 5, 24, 40, 50, 53, 65, 83, 89, 148
plastics, 68
play, 22, 72, 103, 105, 107, 115, 138, 150
plea agreement, 18
plurality, 16
polar bears, 22
pollutant, 40, 101, 102, 138, 140, 141, 142, 145, 162
pollutants, vii, viii, 2, 6, 7, 15, 41, 54, 59, 63, 64, 84, 88, 97, 102, 109, 110, 112, 134, 144, 145, 146, 147, 148, 149, 152, 155
polluters, vii, viii, ix, 33, 34, 137, 138, 141

Index 175

polychlorinated biphenyls (PCBs), 147
poor, 138
population, 11, 22, 120
ports, 15, 160, 161
power, 2, 5, 20, 24, 27, 30, 38, 40, 42, 50, 53, 89, 145, 148
power plants, 2, 5, 20, 53, 89, 145, 148
powers, 59, 115
PPA, 53, 55, 70, 77, 80
PPS, 15
precedents, 92
prejudice, 10
premature death, 148
preparedness, 119
president, 16, 17
pressure, 30
prevention, 2, 5, 68, 80, 84, 119, 138, 143, 148
price changes, 38
price index, 83
prices, 83
primacy, 105, 117, 133
printing, 31
private, 28, 30, 108, 155
proactive, 157
probability, 94
probation, 15, 16, 17, 18, 19, 152, 157, 162
problem-solving, 158
procedural rule, 135
producers, 108
production, 22, 108, 134
professional development, 74
profit, 34, 49, 87
property, 7, 17, 24
protection, viii, 4, 9, 12, 23, 24, 74, 79, 85, 87, 93, 107, 115, 123, 152, 155
protocol, 82, 90
protocols, 79, 89, 90, 121
proxy, 13
PSD, 145, 149, 166
public funds, 49, 96
public health, 27, 34, 50, 53, 66, 68, 74, 76, 79, 93, 94, 98, 106, 119, 120, 139, 147, 158
public interest, 3, 119, 124
public service, 4
Public Works Committee, 135
pulse, 12
punishment, 101
purchasing power, 38
PVC, 8

Q

quality assurance, 39

quality control, 91
query, 126

R

radiation, 109
radioactive waste, 134
rain, 64, 148
rainfall, 146
rainwater, 146
range, 4, 24, 27, 80, 98, 99, 102, 108, 127, 160
real estate, 7
real property, 123
real terms, 57, 60, 62, 66, 83
reality, 140
reclamation, 4
recognition, 30, 93, 143
recognized tribe, 29
record keeping, 66
recovery, 10
recreation, 12
recreational, 156
recycling, 31
redevelopment, 157
refineries, 3, 5, 53, 68, 108
refining, 3, 5, 68
reflection, 112
registry, 109, 113, 126
regulation, 13, 20, 21, 22, 54, 64, 113, 118
regulations, viii, ix, 17, 25, 29, 34, 53, 60, 64, 66, 72, 76, 85, 93, 97, 100, 103, 104, 105, 108, 110, 112, 114, 115, 124, 135, 138, 142
regulators, ix, 16, 58, 59, 65, 76, 93, 97, 101, 103, 112, 113, 119, 125
regulatory requirements, 98, 102, 108, 112, 114, 127
rehabilitate, 156
reimbursement, 2, 9, 141
relationship, 52, 57, 59, 60, 67, 69, 70, 72, 77, 107
relationships, 55, 67, 80, 99, 127
reliability, 35, 49, 57, 78, 92, 96
religion, 14
remediation, 17, 27, 153
remote sensing, 113
repair, 121
reservation, 28, 83
reserves, 28
reservoir, 4, 23, 160
reservoirs, 5, 12
residential, 24, 120
resolution, 21, 160
resource allocation, 52, 56, 67, 81, 94
Resource Conservation and Recovery Act, 8, 16, 17, 21, 22, 27, 53, 63, 64, 66, 67,68, 69, 77, 81, 83,

102, 106, 109, 110, 118, 126, 127, 129, 131, 134, 146, 153, 156, 165, 166

resource management, 13

respiratory, 147, 148

respiratory failure, 148

respiratory problems, 147, 148

responsiveness, 143

restitution, 17, 18, 19, 97, 130, 142, 151, 152, 153

retirement, 50

returns, 114

revenue, 29, 70

Rhode Island, 31, 81, 84, 138

rights-of-way, 22

rings, 19

risk, 17, 20, 21, 106, 158

risk assessment, 21

risks, viii, 4, 21, 27, 34, 41, 68, 102, 120, 147

rivers, 4, 152

rubber, 109

runoff, 64, 68, 146, 148

rural, 138

rural communities, 138

S

S&T, 123

Safe Drinking Water Act (SDWA), 68, 77, 102, 103, 105, 110, 115, 117, 126, 129, 131, 133, 134, 135, 159, 164

safety, 20, 22, 29, 105, 106, 151, 159, 161

sample, 77

sampling, 82, 113

sanctions, 103, 111, 113, 115, 119, 121, 125, 135

savings, 118

school, 9, 17, 21, 68, 107, 147, 148

search, 23, 127

search engine, 127

searches, 126

searching, 23

Securities and Exchange Commission (SEC), 104

security, 3, 26, 27, 30, 116, 134

sediment, 2, 6, 7, 146, 148

sediments, 21, 156

seismic, 22

selecting, 67, 69, 105

self-monitoring, 112, 144

self-report, 92, 111, 117

semi-structured interviews, 56

senate, 132

Senate, 53, 119, 132, 135, 136

sensing, 113

sentences, 142, 153

sentencing, 18

September 11, 21

services, 30, 31, 159

SES, 135

settlements, viii, 1, 3, 4, 5, 6, 7, 9, 10, 24, 41, 42, 68, 97, 108, 111, 114, 119, 138

severity, 111

sewage, 3, 6, 16, 53, 146, 162

shape, 25

sharing, 30, 121, 161

shellfish, 157

shipping, 15, 16, 27

short-term, 50

sign, 4, 23, 30, 41

signs, 18

simple linear regression, 50

sites, 6, 8, 9, 10, 30, 66, 109, 110, 113, 126, 134, 138, 141, 142, 155, 156, 157

Small Business Administration, 87, 96, 143

smelting, 9

smog, 148

smugglers, 26

smuggling, 19

SO2, 2, 5, 134, 145, 148

socioeconomic, 120

sodium, 16

software, 30

soil, 16, 141, 146, 147, 153, 155, 156

soils, 147

solid waste, 67, 83, 84, 103, 107, 113, 143, 147, 159

Solid Waste Disposal Act, 53, 63, 83, 102

solvents, 147, 148, 159

South Africa, 25, 161

South Carolina, 77, 81, 158

South Dakota, 79

species, 3, 4, 11, 12, 13, 14, 19, 23, 24, 28

specificity, 54

spectrum, 104

spills, 3, 6, 8, 147, 152, 156

staffing, 31, 52, 57, 58, 60, 92, 100, 116, 124

stages, 29, 58, 71, 73, 94, 103, 104

stakeholder, 103

stakeholder groups, 103

stakeholders, 23, 57, 69, 81, 101, 120

standards, 5, 13, 16, 35, 41, 55, 57, 63, 65, 66, 78, 83, 84, 94, 102, 103, 105, 114, 134, 146, 147, 159, 161

State Department, 53, 70

state laws, 75

state office, 72

state oversight, 9, 56, 78, 89, 93

state planning, 100

state regulators, ix, 58, 59, 65, 76, 97

statistical analysis, 34

Index

statistics, 125

statute of limitations, 13, 30

statutes, vii, 1, 8, 34, 41, 49, 53, 54, 60, 76, 93, 98, 99, 100, 101, 102, 104, 105, 106, 107, 108, 109, 110, 111, 113, 114, 115, 116, 117, 118, 121, 123, 124, 126, 133, 135, 140, 151, 166

statutory, ix, 10, 26, 28, 36, 57, 69, 92, 97, 98, 100, 101, 102, 103, 106, 110, 112, 114, 122, 124, 127

statutory provisions, 103, 106

storage, 16, 54, 66, 67, 68, 118, 152, 155, 159

stormwater, 142, 146, 152

strains, 72, 107

strategic planning, 94, 95, 103, 106, 109, 132

strategies, 65, 73, 86, 91, 92, 93, 112, 120, 140

streams, 161

strength, ix, 137

submarines, 134

substances, 8, 98, 120

suburbs, 23

sulfur, 134, 145, 147

sulfur dioxide, 134, 145

sulfur oxides, 147

summaries, 126

summer, 67

superfund, 2, 8, 9, 10, 98, 102, 105, 110, 117, 123, 126, 131, 135, 138, 141, 153, 154, 155, 156, 165, 166

supervised release, 153

supplemental, viii, 5, 6, 7, 40, 97, 111, 119

supply, 115, 160

support services, 31

Supreme Court, 3, 13, 16, 25, 26, 41, 118

surface water, 54, 146, 147, 156

symptoms, 148

T

tanks, 16, 18, 66, 67, 68, 84, 152, 153, 159

targets, 64

task force, 115

taxpayers, 24

technical assistance, 54, 120, 160

telecommunications, 30

telecommunications services, 30

Tennessee, 151

territorial, 83, 132, 133

testimony, viii, 74, 75, 85, 86, 87, 96

Texas, 23, 26, 31, 149, 152

Thailand, 25

threat, 14, 107, 112

threatened, 3, 4, 11, 23, 28, 64, 152

threats, 11, 13, 147

thresholds, 81, 84

tiger, 11

timber, 13, 24, 25, 29

time frame, 35, 80, 113

time series, 43

timing, 54

title, 27, 63, 65

Title III, 65, 83

top management, 95

tort, 10, 27

total costs, 63

toxic, 41, 65, 98, 120, 127, 146

toxic substances, 98, 120

Toxic Substances Control Act, 102, 105, 126, 127, 129, 131, 134

toxicity, 41

tracking, 30, 109, 112

trade, 23, 25

trading, 101, 102

traffic, 22

training, 3, 25, 26, 27, 30, 31, 54, 57, 69, 72, 74, 104, 116, 134, 138, 142, 160, 161

training programs, 74

transaction costs, 150

transfer, 70

transmission, 14, 22

transparency, 34, 38, 40, 42, 43, 92

transport, 19, 143

transportation, 16, 18, 108

Treasury, 35, 38, 117, 135

treaties, 107

trial, 8, 9, 10, 14, 17, 18, 26, 29, 30, 31, 41

tribal, 4, 28, 29, 67, 79, 98, 99, 103, 107, 108, 127, 133, 144, 145, 147, 159, 166

tribal lands, 107, 159

tribes, 4, 14, 28, 29, 54, 62, 63, 100, 103, 104, 107, 113, 120, 133, 147, 159

trichloroethylene, 159

triggers, 157

trucking, 27

trust, vii, 1, 4, 7, 28, 29, 66

trust fund, 66

tundra, 18

turtle, 19

turtles, 19

U

U.S. Treasury, 35, 38, 117, 135

uncertainty, 13, 42

Underground Storage Tank Trust Fund, 66, 84

uniform, 62, 73, 75, 79, 81

United Kingdom, 25

United Nations, 15

Index

United States, v, 1, 2, 3, 4, 5, 6, 7, 8, 9, 10, 11, 12, 13, 14, 15, 16, 17, 18, 19, 21, 23, 24, 25, 26, 27, 28, 29, 30, 41, 42, 53, 83, 98, 107, 123, 126, 134, 135, 137, 138, 158
universe, 80, 92, 98, 108, 109, 110
universities, 121, 138, 143, 158
uranium, 10
urban areas, 146
USDA, 13
Utah, 24, 27, 56, 69, 70, 77, 79

V

values, 38, 49, 83, 96, 101
variability, 100, 105, 113, 124, 127
variation, 21, 74, 75, 79, 86, 87, 89, 91
vegetation, 156
vehicles, 108, 156
vermiculite, 17
Vermont, 84, 138
vessels, 3, 15, 153
Vietnam, 25
Vietnamese, 19
vinyl chloride, 8
visible, 134

W

warfare, 3, 26
waste disposal, 21, 143
waste incinerator, 27
waste management, 66, 146
wastes, 16, 17, 22, 66, 109, 146, 153, 157
wastewater, 2, 6, 7, 15, 108, 138, 146, 148, 149

wastewater treatment, 6
water quality, 4, 7, 23, 64, 94, 146, 148
water rights, 4, 23, 28
water supplies, 146
waterways, 2, 3, 5, 6, 7, 15, 18, 146, 148
weakness, 73
weapons, 27, 134
wells, 156
wetlands, 7, 16, 23, 105, 160
wilderness, 12
wildfire, 4, 13
wildlife, vii, 1, 4, 19, 22, 24, 25, 123, 146, 156
wildlife conservation, 24
wind, 160
wind turbines, 160
Wisconsin, 29
withdrawal, 79
witnesses, 24, 30
workers, 17, 153, 162
workforce, 55, 60, 62, 86, 88, 90, 92, 93, 95, 140
workload, 52, 55, 56, 57, 58, 60, 64, 65, 66, 72, 78, 83, 92, 95
World Trade Center, 21
World War II, 10
Wyoming, 79, 160

X

X-rays, 153

Y

yield, 11, 73, 82